Dynamics of Statistical Experiments

Series Editor
Nikolaos Limnios

Dynamics of Statistical Experiments

Dmitri Koroliouk

iSTE

WILEY

First published 2020 in Great Britain and the United States by ISTE Ltd and John Wiley & Sons, Inc.

ISTE Ltd
27-37 St George's Road
London SW19 4EU
UK

www.iste.co.uk

John Wiley & Sons, Inc.
111 River Street
Hoboken, NJ 07030
USA

www.wiley.com

Library of Congress Control Number: 2019956286

British Library Cataloguing-in-Publication Data
A CIP record for this book is available from the British Library
ISBN 978-1-78630-598-5

MIX
Paper from responsible sources
FSC® C013604

Contents

Preface

This monograph is based on the Wright–Fisher model (Ethier and Kurtz 1986, Ch. 10) in mathematical theory of population genetics, considered as a dynamical experimental data flow, expressed in terms of the fluctuations, which are deviations from a certain equilibrium point (steady-state).

Statistical experiments (SEs) are a stochastic point of view of the collective behavior of a finite number of interacting agents taking a finite number of decisions $M \geq 1$ ($M = 1$ corresponds to the basic binary model with two alternatives).

The dynamics, in discrete time, is described for *stochastic difference equations* (SDEs) that consist of two components: evolutionary processes (EPs) (the predictable component) and martingale differences (the stochastic processes).

An essential feature of SEs is that their characterization by *regression function of increments* (RFI), defines the EP (drift of SEs). The basic assumption for binary SEs is given by linear fluctuations with respect to an equilibrium point (steady state). The RFI can be interpreted as a fundamental principle "stimulation and deterrence": at each stage, the EP increments decrease proportionally to the current fluctuation value.

Linear fluctuations of the RFI are basic models in the diffusion approximation of SEs in discrete–continuous time (Chapter 2).

The dynamics of SE, in discrete time, provides an effective statistical estimation of a linear drift factor V_0 under the assumption of stationarity (section 3.1).

Stationary Gaussian Markov SEs, given by the solution of a SDE of increments, are characterized by two-dimensional covariance matrices. The expression of their dynamics in terms of the linear fluctuations with respect to an equilibrium value gives two equivalent determinations of the SEs: as the solution of the SDEs and by two-dimensional covariance matrices (Theorem 3.2.3).

The collective behavior of a finite number of N agents is given by the increment of SEs in a SDE containing two components: an evolutionary process (a predictable component) that determines the fluctuations of a given state with respect to the equilibrium and also martingale differences (a stochastic component) approximated by a normal Brownian motion as $N \to \infty$.

The problem of statistical hypothesis verification is formulated in the classification of EPs, which determine the dynamics of the predictable component. The main classification of EPs reflects the trajectories behavior in the neighborhood of the equilibrium point and is subdivided into two types of behavior: attractive and repulsive.

Statistical experiment models are a valuable tool for advanced graduate students and practicing professionals in statistical modeling and dynamics of complex systems consisting of a large number of interacting elements.

Dmitri KOROLIOUK
December 2019

List of Abbreviations

AF	action functional
ASD	asymtotically small diffusion
DEE	difference evolutionary equation
DMD	discrete Markov diffusion
DMP	discrete Markov process
FCS	fluorescence correlation spectroscopy
EG	exponential generator
EP	evolutionary process
ESE	exponential statistical experiment
MC	Markov chain
MP	Markov process
MRE	Markov random environment
MSE	multivariate statistical experiment
OEF	optimal estimating function
QSF	quasi-score functional
ROI	region of interest
RF	regression function
RFI	regression function of increments
SA	stochastic approximation
SDE	stochastic difference equation
SE	statistical experiment
TSE	ternary statistical experiment

Introduction

The main objective of consideration is a *statistical experiment* (SE), defined as the averaged sum $S_N(\cdot)$ of independent and identically distributed random samples, which take a finite number of values, in particular, the binary values ± 1:

$$S_N^{\pm}(k) := \frac{1}{N}\sum_{n=1}^{N} \delta_n^{\pm}(k), \quad \delta_n^{\pm}(k) := I\{\delta_n(k) = \pm 1\}, \quad k \geq 0.$$

The basic assumption 3 (Proposition 1.3.3) is derived by using the *regression function of increments* (RFI), given by the predictable component of SEs [1.3.10].

SE presentation as averaged sums of independent and identically distributed random samples (which take a finite number of values) means that an SE is defined by two components: evolutionary processes (EPs) (predictable components) and martingale differences (stochastic components). So, the SE can be considered as special semi-martingales (Jacod and Shiryaev 1987).

An essential feature of specifying SEs is their characterization by *stochastic difference equations* (SDEs) consisting of two parts: a predictable component defined by RFI and a stochastic component characterized by its first two moments.

Proposition 1.2.3 (*Basic assumption 3*). The SEs given by the averaged sums of the sample values are determined by the solutions of the SDEs:

$$\Delta S_N^{\pm}(k+1) = -V_0(S_N^{\pm}(k)) \cdot [S_N^{\pm}(k) - \rho_{\pm}] + \Delta\mu_N^{\pm}(k+1), \quad k \geq 0.$$

The stochastic component of the SDEs [1.2.34] and [1.2.35] is characterized by their quadratic characteristics.

The existence of equilibria $\rho_{\pm}$ of the predictable component provides the convergence with probability 1 (Theorem 1.2.2).

However, the stochastic component defined by the martingale differences generates, as $N \to \infty$, a series (by k) of normally distributed random variables with certain quadratic variations which depend on the state of SEs (Theorem 1.2.3). Asymptotic representation of the normalized martingale differences provides a basic stochastic approximation by SDEs, with the stochastic part represented by the normally distributed martingale differences (Propositions 1.2.4 and 1.2.5).

The multivariate statistical experiments (MSEs) in section 1.3 are considered in the assumption of a finite number of the possible values $E = \{e_0, e_1, \dots, e_M\}$, $M \geq 1$. The linear RFI is determined as follows.

Proposition 1.3.1 (*Basic assumption 1*)

$$\Delta P_m(k+1) = -V_0[P_m(k) - \rho_m], \quad 0 \leq m \leq M, \quad k \geq 0, \quad V_0 > 0.$$

That is, the linear RFI is given by the fluctuation of increments with respect to the equilibrium value.

The essential result is given in *Proposition 1.3.2* where the multivariate frequencies with Wright–Fisher normalization are considered for models in population genetics (Ethier and Kurtz 1986). The nonlinear RFI are represented as in canonical representation [1.3.9] with the fluctuation of increments with respect to the ratio of the values of EPs and the corresponding equilibria.

Proposition 1.3.3. (*Basic assumption 3.*) The frequencies of multivariate EPs are determined by *difference evolutionary equations* (DEEs) with nonlinear RFIs having the fluctuation representation [1.3.10].

The presence of the equilibrium state ensures the convergence of EPs, as $k \to \infty$ (Theorem 1.3.1).

The SDE for MSEs is given with martingale difference [1.3.16] as the stochastic component with quadratic characteristics (Lemma 1.3.1; also see Proposition 1.3.4). A representation of MSEs as normalized sums of independent and identically distributed random variables [1.3.28] provides the convergence with probability 1 (Theorem 1.3.2).

In sections 1.3.1 and 1.3.2, the martingale differences are approximated by normally distributed random variables (Theorem 1.3.3). The MSEs are determined by the solutions of the SDEs with normally distributed stochastic components and the "constant" quadratic characteristics defined by equilibrium points (Proposition 1.3.5).

The SEs with Wright–Fisher normalization are developed with RFIs given by the fluctuations with respect to equilibrium introduced in section 1.4. The binary EPs are transformed in the increment probabilities with RFIs in terms of the fluctuations [1.4.12]. In this connection, the RFIs are postulated as basic assumption 2 (Proposition 1.4.1).

Section 1.5 is dedicated to the *exponential SE* (ESE) defined as a product of random variables (Definition 1.5.1; also see [1.5.7]).

The ESE is investigated in two normalized schemes: Theorem 1.5.1—the convergence, in probability, of ESEs with normalization $\lambda_N = \lambda/N$, $N \to \infty$, is realized using Le Cam's approximation (Borovskikh and Korolyuk 1997) and Theorem 1.5.2—the convergence, in distribution, of ESEs with normalization $\lambda_N = \lambda/\sqrt{N}$, $N \to \infty$, to geometric Brownian motion.

Theorems 1.5.1 and 1.5.2 give us an opportunity to represent ESEs in exponential approximation scheme [1.5.41] using only three parameters for the normal process of autoregression. Note that the ESE has important interpretation in financial mathematics (Shiryaev 1999).

Chapter 2 is dedicated to the diffusion approximation of SEs in discrete–continuous time.

Discrete Markov processes (DMPs), determined by the solutions of the SDEs in discrete–continuous time, are approximated, as $N \to \infty$, by diffusion processes with evolution, given by *differential stochastic equations*.

The discrete–continuous time is determined by the connection of discrete instants of time $k \geq 0$, $k \in \mathbb{N} = \{0, 1, \ldots\}$ with continuous time $t \geq 0$, $t \in R_+ = [0, +\infty)$ by the formula $k = [Nt]$, $t \geq 0$. The integer part $[Nt] = k$

defines a discrete sequence $t_k^N := k/N$, $k \geq 0$. The adjacent moments of the continuous time $\Delta_N := t_{k+1}^N - t_k^N = 1/N \to 0$ and $t_k^N \to t$ as $N \to \infty$, $k \to \infty$.

DMPs in discrete–continuous time are considered with the normalized fluctuation [2.1.12]. Basic assumption 2.1.2 supposes that the DMP, given by the solution of the SDE [2.1.15] with nonlinear RFIs, can be approximated by a linear SDE [2.1.20] (see basic assumption 2.1.3). The finite-dimensional distributions of DMPs converge, as $N \to \infty$, to a diffusion process with evolution of Ornstein–Uhlenbeck type given by the solution of the SDE [2.1.24] (Theorem 2.2.1, Conclusion 2.2.1).

The diffusion approximation of DMPs is based on the operator characterization of Markov processes (Korolyuk and Limnios 2005).

Propositions 2.2.1 and 2.2.2 are determined the SDEs with predictable components, as well as the quadratic characteristics of stochastic component convergence, as $N \to \infty$, to equilibrium values (Lemma 2.2.1) provide the linear character of the SDEs for limit processes, characterized only by the local fluctuations (Proposition 2.2.3).

The DMP in discrete and continuous Markov random environments (MREs) are considered in section 2.3. Using the method of singular perturbation (Korolyuk and Limnios 2005), Theorem 2.3.1 states the convergence of the finite-dimensional distributions to those of the limit diffusion processes with evolution, defined by averaged drift and diffusion parameters $\widehat{V}$ and $\widehat{\sigma}^2$ [2.3.10], respectively. The averages are intend over the stationary distribution of the embedded Markov chain. Theorem 2.3.2 states the limit diffusion process with evolution, with parameters $\widehat{V}$ and $\widehat{\sigma}^2$ averaged by the stationary distribution of a Markov ergodic environment.

Essential applications can give the technique of singular perturbation problem for a reducible invertible operator acting on a perturbed test function $\varphi_1(c, x)$ that provides the solvability condition of a certain operator equation (see [2.3.38]–[2.3.39]).

One specific model in section 2.4 is DMP in a balanced Markov random environment. The approximation of DMP in discrete–continuous time with balance condition [2.4.9] is given in Theorem 2.4.1. The limit Ornstein–Uhlenbeck diffusion process [2.4.20] is determined by the parameters $\widehat{V}_0$ and

$\widehat{\sigma}_0^2$ calculated by the explicit formulas [2.4.15]–[2.4.16]. Here the solution of a singular perturbation problem for the truncated operator [2.4.35] is realized on the perturbed test function [2.4.36] that consists of three differently scaled components.

In section 2.5, the adapted SEs are combined with a random time change, which transforms a discrete stochastic basis $\mathfrak{B}_{\mathbb{N}} = (\Omega, \mathfrak{F}, (\mathfrak{F}_k, k \in \mathbb{N}), \mathcal{P})$ into a continuous stochastic basis $\mathfrak{B}_T = (\Omega, \mathfrak{G}, (\mathfrak{G}_t, t \in \mathbb{R}_+), \mathcal{P})$.

The adapted SE with continuous basis $\mathfrak{B}_T$ are studied in the series scheme with a small series parameter $\varepsilon \to 0$ ($\varepsilon > 0$). The limit diffusion process with evolution is determined in Theorem 2.5.1 by the predictable characteristics $\widehat{V}_0$ and $\widehat{\sigma}_0^2$.

In section 2.6, the DMPs are considered in an asymptotic diffusion environment, generated by the DEE [2.6.6] with the balance condition [2.6.7]. Theorem 2.6.1 states the convergence by $\varepsilon \to 0$ to an Ornstein–Uhlenbeck diffusion process.

In section 2.7, the DMPs with asymptotically small diffusion are considered. Its exponential generator is determined in Theorem 2.7.2 by relation [2.7.2] on the test functions $\varphi(c) \in C^3(R)$. The action functional for DMPs is introduced, following the method developed in monograph (Freidlin and Ventzell 2012).

Chapter 3 is devoted to statistically estimating the drift parameter V_0 and verifying the hypothesis for the SE dynamics as $k \to \infty$.

The case of one-dimensional stationary SEs is considered in section 3.1. The predictable component is supposed to have a linear form with drift parameter $V_0 > 0$.

Theorem 3.1.1 establishes the wide-sense stationarity conditions of SEs $(\alpha_t, t \geq 0)$, given by the solution of the SDEs

$$\Delta\alpha_t = -V_0\alpha_t + \Delta\mu_t, \quad t \geq 0, \quad 0 < V_0 < 2, \qquad [0.0.1]$$

as a relation between the dispersion of stationary SE $E\alpha_t^2 = \sigma_0^2$ and the quadratic characteristic of the stochastic component:

$$\sigma^2 = \sigma_0^2 \cdot \mathcal{E}^0, \quad \sigma^2 := E[(\Delta\mu_t)^2], \quad \mathcal{E}^0 := V_0(2 - V_0). \qquad [0.0.2]$$

A wide-sense stationarity, SE α_t, $t \geq 0$, given by the solution of the SDE [0.0.1], is characterized by the covariance matrix

$$\mathbb{R} = \begin{bmatrix} R & R^0 \\ R^0 & R^\Delta \end{bmatrix} \qquad [0.0.3]$$

of two-component vector $(\alpha_t, \Delta\alpha_t)$, $t \geq 0$, given by the following relations (Theorem 3.1.2):

$$\begin{gathered} R := E\alpha_t^2 = \sigma_0^2, \quad R^\Delta := E[\Delta\alpha_t]^2 = 2V_0\sigma_0^2, \\ R^0 := E[\alpha_t \Delta\alpha_t] = -V_0\sigma_0^2. \end{gathered} \qquad [0.0.4]$$

The additional condition on a stationary SE can be supposed to be a Gaussian distribution of the stochastic component. Supposing the normal distribution of the initial value α_0: $E\alpha_0 = 0$, $E\alpha_0^2 = \sigma_0^2$, the matrix of the quadratic form R^{-1} which generates two-dimensional normal distribution and can be represented by elements of the covariance matrix [3.1.9] follows the relation $\sigma^2 = \sigma_0^2 \cdot 2V_0W$, $W := 1 - V_0/2$. In Corollaries 3.1.3 and 3.1.4, formulae [3.1.21] and [3.1.22] determine the drift parameter V_0 and the correlation coefficient r^2.

The trajectories of a stationary SE generate the covariance statistics

$$R_T := \frac{1}{T}\sum_{t=0}^{T-1} \alpha_t^2, \;\; R_T^0 = \frac{1}{T}\sum_{t=0}^{T-1} \alpha_t \Delta\alpha_{t+1}, \;\; R_T^\Delta := \frac{1}{T}\sum_{t=0}^{T-1} (\Delta\alpha_{t+1})^2,$$

which serve, by virtue of relations [0.0.4], as the basis of consistent estimates of the drift parameter V_0:

$$V_T = -\frac{R_T^0}{R_T}, \quad V_T^\Delta = \frac{R_T^\Delta}{2R_T}, \quad T > 0, \qquad [0.0.5]$$

each of which estimate the drift parameter V_0.

Theorem 3.1.3 states the convergence with probability 1:

$$V_T \xrightarrow{P1} V_0, \quad T \to \infty.$$

The main statistical problem in analysis of stationary SEs is to construct the *optimal estimating function* (OEF) for justifying the statistical estimation of the drift parameter V_0. The OEF is given by the quadratic variation (Heyde 1997) of the martingale differences:

$$M_T(V, V_0) := \frac{1}{T}\sum_{t=0}^{T-1}[\Delta\alpha_t + V\alpha_t]^2 = V^2 R_T + 2V R_T^0 + R_T^\Delta.$$

The statistics of one-dimensional stationary SEs are also generalized to the multivariate stationary case (section 3.2).

Under some natural conditions of stationarity, in wide-sense, [3.2.3] is established in Theorem 3.2.1. The positive-definite covariance matrix $\mathbb{R} := E[\alpha_t\alpha_t']$, $t \geq 0$, is given by the solution of the matrix equation

$$\mathbb{B} = \mathbb{V}_0\mathbb{R} + \mathbb{R}\mathbb{V}_0{}' - \mathbb{V}_0\mathbb{R}\mathbb{V}_0{}', \ \ \mathbb{B} := E[\Delta\mu_t\,\Delta\mu_t{}']. \qquad [0.0.6]$$

Representation of the OEF in Theorem 3.2.1 gives the following *a priori* statistic:

$$\mathbb{V}_T^0 = -\mathbb{R}_T^0\mathbb{R}_T^{-1}.$$

The residual term $\mathcal{B}_T$ in the representation

$$\mathcal{E}_T(\mathbb{V}, \mathbb{V}_0) = \mathcal{E}_T^0(\mathbb{V}, \mathbb{V}_T^0) + \mathcal{B}_T, \quad T > 0,$$

gives the estimator of the stationary factor $\mathcal{E}^0$ (see [0.0.2]).

In section 3.2.3, using the normal correlation theorem (Liptser and Shiryaev 2001, Ch. 13), any stationary Gaussian Markov SE, defined by its two-component covariances, is determined by the solution of the SDE [3.2.27] with Gaussian martingale differences [3.2.28].

Section 3.4 is dedicated to asymptotic analysis of EP dynamics determined by the solution of the DEE [3.4.2] or [3.4.8]. The classification of EPs is based on their trajectories, taking into account the equilibrium values $\rho_\pm$ as well as the values of the drift parameter V_0.

The classification of frequency EPs is given in Propositions 3.4.1 and 3.4.2. Its justification is based on the limit behavior of frequency EPs, as $k \to \infty$. In Theorem 3.4.1, the probabilities of alternatives $P_\pm(k)$, $k \geq 0$, given by the DEE solution [3.4.2] or [3.4.8], have four possible asymptotic behavioral trends: attractive $\lim_{k\to\infty} P_{\pm}(k) = \rho_{\pm}$; repulsive $\lim_{k\to\infty} P_{\pm}(k) = 1$ and $\lim_{k\to\infty} P_{\mp}(k) = 0$ under initial condition $P_{\pm}(0) > \rho_{\pm}$ or $P_{\pm}(0) < \rho_{\pm}$; domination $\pm$ $\lim_{k\to\infty} P_{\pm}(k) = 1$ and $\lim_{k\to\infty} P_{\mp}(k) = 0$.

The interpretation of the limit behavior of SEs in practical applications such as population genetics (Crow and Kimura 2009; see also Fisher 1930, Schoen 2006, Wright 1969), economics and finance (Shiryaev 1999) and others have a certain practical interest. Another interpretation of SE asymptotic trends is given for the models of collective behavior or collective teaching (section 3.4.6).

The convergence of the SDEs solutions in a stochastic approximation scheme (Nevelson and Has'minskii 1976) is proved using the classifiers defined by truncated RFIs (Definition 3.5.1).

Classification of SEs is based on limit theorems for the stochastic approximation of SEs (Theorem 3.5.2).

Sections 3.6 and 3.7 give an important illustrative example of MSEs with three possible decisional alternatives. The classification theorems are considered for EPs as well as for the ternary SEs.

In Chapter 4, we use the mathematical model of *discrete Markov diffusion* (DMD) to describe the mechanisms of interaction of biological macromolecules. This model is determined by the solution of the SDEs with predictable and stochastic components.

The basic statistical data is derived from the dynamic monitoring of macromolecules (Rigler and Elson 2001) in *fluorescence correlation spectroscopy* (FCS), which calculates the rate of fluctuations of fluorescence-labeled molecules in a confocal space of observations.

For this purpose, we use the estimates, by trajectories, of the DMD models' statistical parameters: V and σ^2 for mathematical description of mechanisms in collective biological interactions.

The mathematical model and its verification on the simulated data set are obtained on the basis of the well-known Stokes–Einstein model. In particular, they used numerically generated data of a mixture of particles with two values of the diffusion coefficient: $D_1 = 10$ and $D_2 = 100\,\mu\text{m}^2$. Such simulated data, considered as trajectories of the new mathematical model of the DMD, has shown good discriminatory properties for the revelation of a mixture of two Brownian movements: “fast” and “slow.”

Our hypothesis is based on two different considerations: the physical (Stokes–Einstein) diffusion process and the mathematical (stationary discrete Markov) diffusion process are different models that describe the same physical phenomenon.

In addition, in analyzing data for mixing particles with different diffusion coefficients, the theoretical parameters of the model V (drift parameter) and σ^2 (squared dispersion of the stochastic component) have almost linear discriminatory ability to the molar determination of the mixture of fractions.

The proposed method of statistical analysis of real measurements is especially important because it is specially designed for the analysis of microfluorescence measurements in an area where the observation of physical data (biological, chemical, etc.) has no basis in the form of a simple physical theory (e.g., Koroliouk *et al.* 2016b).

1

Statistical Experiments

1.1. Statistical experiments with linear regression

Statistical experiments (SEs) are defined by stochastic difference equations (SDEs; section 1.1.4). Evolutionary processes (EPs) are defined by difference evolutionary equations (DEEs), with a linear regression function of increments (RFI; section 1.1.2). The dynamics of EPs is characterized by the fundamental principle "stimulation and deterrence." We establish the convergence of SEs to the equilibrium state (section 1.1.5) and the normal approximation of the stochastic component (section 1.1.6) as the sample volume $N \to \infty$.

1.1.1. *Basic definitions*

Binary SEs are given by averaged sums of random samples $\delta_n(k)$, $1 \leq n \leq N$, $k \geq 0$, jointly independent at fixed $k \geq 0$ and equally distributed at different $n \in [1, N]$, which take two values ± 1:

$$S_N(k) := \frac{1}{N} \sum_{n=1}^{N} \delta_n(k), \quad -1 \leq S_N(k) \leq 1, \quad k \geq 0. \qquad [1.1.1]$$

The binary SEs have a representation of the frequency differences:

$$S_N(k) = S_N^+(k) - S_N^-(k), \quad k \geq 0, \qquad [1.1.2]$$

$$S_N^{\pm}(k) := \frac{1}{N} \sum_{n=1}^{N} \delta_n^{\pm}(k), \quad \delta_n^{\pm}(k) := I\{\delta_n(k) = \pm 1\}, \quad k \geq 0. \qquad [1.1.3]$$

Here, as usual, $I(A)$ is the indicator of a random event A, and the value $S_N^{\pm}(k)$ specifies the frequency SEs (positive or negative, resp.) in the kth sample.

Note that there is an obvious identity

$$S_N^{+}(k) + S_N^{-}(k) \equiv 1, \quad k \geq 0. \qquad [1.1.4]$$

So, the binary SEs [1.1.1] and [1.1.2] unambiguously determine the frequency SEs and vice versa:

$$\begin{aligned} S_N^{\pm}(k) &= \tfrac{[1 \pm S_N(k)]}{2}, \\ S_N(k) &= 2S_N^{+}(k) - 1 = 1 - 2S_N^{-}(k). \end{aligned} \qquad [1.1.5]$$

DEFINITION 1.1.1.– *The predictable components* of SEs [1.1.1] and [1.1.3] are determined by conditional expectations:

$$P_{\pm}(k+1) := E[\delta_n^{\pm}(k+1) \mid S_N^{\pm}(k) = P_{\pm}(k)], \; 1 \leq n \leq N, \; k \geq 0, \qquad [1.1.6]$$

$$C(k+1) := E[\delta_n(k+1) \mid S_N(k) = C(k)], \quad 1 \leq n \leq N, \quad k \geq 0, \qquad [1.1.7]$$

for all n, where $1 \leq n \leq N$.

The predictable component of the SE [1.1.6] determines the frequency probabilities of the choice indicators of alternatives $\delta_n^{\pm}(k+1)$ on the next $(k+1)$th stage under the condition of a given SE state on the previous kth stage $S_N^{\pm}(k) = P_{\pm}(k)$.

This is the *statistical principle* manifestation in the dynamics of an alternative choice in the presence of information about an SE average state in the previous stage.

DEFINITION 1.1.2.– EPs are determined by conditional expectations

frequency: $P_{\pm}(k+1) = E[S_N^{\pm}(k+1) \mid S_N^{\pm}(k) = P_{\pm}(k)], \quad k \geq 0,$ [1.1.8]

binary: $C(k+1) = E[S_N(k+1) \mid S_N(k) = C(k)], \quad k \geq 0.$ [1.1.9]

The EPs [1.1.8] and [1.1.9] do not depend on the sample size N.

Evidently, the EPs [1.1.8] and [1.1.9] satisfy the relations

$$\begin{aligned} C(k) &= P_{+}(k) - P_{-}(k), \quad k \geq 0, \\ P_{+}(k) + P_{-}(k) &\equiv 1 \quad \forall k \geq 0. \end{aligned} \qquad [1.1.10]$$

Relations [1.1.10] allow us to reduce the analysis of the frequency EPs to the analysis of the binary EPs using the following formula:

$$P_{\pm}(k) = \frac{1}{2}[1 \pm C(k)], \quad k \geq 0. \qquad [1.1.11]$$

The dynamics of EPs [1.1.8] and [1.1.9] are determined for the increments

$$\begin{aligned} \Delta P_{\pm}(k+1) &:= P_{\pm}(k+1) - P_{\pm}(k), \quad k \geq 0, \\ \Delta C(k+1) &:= C(k+1) - C(k), \quad k \geq 0. \end{aligned} \qquad [1.1.12]$$

The increments of binary EPs are determined by differences between the frequency evolution processes:

$$\Delta C(k+1) = \Delta P_{+}(k+1) - \Delta P_{-}(k+1), \quad k \geq 0. \qquad [1.1.13]$$

The increments of frequency EPs are determined by conditional mathematical expectations:

$$V_0^{\pm}(P_{\pm}(k)) := E[\Delta S_N^{\pm}(k+1) \mid S_N^{\pm}(k) = P_{\pm}(k)]. \qquad [1.1.14]$$

Similarly, the increments of binary EPs are determined by conditional mathematical expectations:

$$V_0(C(k)) := E[\Delta S_N(k+1) \mid S_N(k) = C(k)]. \qquad [1.1.15]$$

Obviously, we obtain the following relations:

$$\begin{aligned} &V_0(C(k)) = V_0^{+}(P_{+}(k)) - V_0^{-}(P_{-}(k)), \\ &V_0^{\pm}(P_{\pm}(k)) = \tfrac{1}{2}[1 \pm V_0(C(k))]. \end{aligned} \qquad [1.1.16]$$

In addition, the following global balance condition is fulfilled:

$$V_0^{+}(p_{+}) + V_0^{-}(p_{-}) = 0, \quad p_{+} + p_{-} = 1.$$

1.1.2. *Difference evolution equations*

The dynamics of EPs given by conditional mathematical expectations [1.1.14] and [1.1.15] are determined by the fluctuations of EPs to the corresponding equilibriums.

PROPOSITION 1.1.1.– (*Basic assumption 1.*) Binary EPs [1.1.9] are determined by the solutions of the difference evolution equations:

$$\Delta C(k+1) = -V_0[C(k) - \rho], \quad k \geq 0, \quad V_0 > 0. \qquad [1.1.17]$$

Frequency evolution processes [1.1.8] are determined by the solutions of the difference evolution equations

$$\Delta P_{\pm}(k+1) = -V_0[P_{\pm}(k) - \rho_{\pm}], \quad k \geq 0, \quad V_0 > 0. \qquad [1.1.18]$$

The parameters ρ and $\rho_{\pm}$ specify the equilibrium of the RFI [1.1.17] and [1.1.18] and satisfy the conditions:

$$0 \leq \rho_{\pm} \leq 1, \quad \rho_+ + \rho_- = 1, \quad \rho := \rho_+ - \rho_-, \quad |\rho| \leq 1. \qquad [1.1.19]$$

The obvious relation

$$\begin{aligned} p_+ - \rho_+ &= -(p_- - \rho_-) = \rho_- p_+ - \rho_+ p_- \\ &= \rho_+ \rho_- (p_+/\rho_+ - p_-/\rho_-) \end{aligned} \qquad [1.1.20]$$

explains *the fundamental principle of the dynamics of EPs,* that is, "stimulation and deterrence." The increment of EPs at each stage decreases in proportion to the value of the fluctuation at this stage with a coefficient of proportionality $V_0 > 0$.

We obtain the following obvious global balance condition:

$$(p_+ - \rho_+) + (p_- - \rho_-) \equiv 0, \quad k \geq 0, \qquad [1.1.21]$$

which ensures the correctness of difference evolution equations [1.1.18].

The presence of the equilibria ρ and $\rho_{\pm}$ in the difference evolutional equations [1.1.17] and [1.1.18] indicates that the increments of the EPs decrease to zero, that is, the value of the EP reaches equilibrium.

1.1.3. *The equilibrium state*

The basic assumption asserts that the frequency RFI is given by the following conditional mathematical expectation:

$$V_0^{\pm}(p_{\pm}) := E[\Delta S_N^{\pm}(k+1) \mid S_N^{\pm}(k) = p_{\pm}], \qquad 0 \leq p_{\pm} \leq 1, \quad p_+ + p_- = 1. \qquad [1.1.22]$$

Similarly, the binary RFI is given by the following conditional mathematical expectation:

$$V_0(c) := E[\Delta S_N(k+1) \mid S_N(k) = c], \quad |c| \leq 1. \qquad [1.1.23]$$

Accordingly, the frequency and binary linear *regression functions* (RFs) are also given by the following conditional mathematical expectations:

$$V^{\pm}(p_{\pm}) = E[S_N^{\pm}(k+1) \mid S_N^{\pm}(k) = p_{\pm}],$$
$$0 \leq p_{\pm} \leq 1, \quad p_+ + p_- = 1. \qquad [1.1.24]$$

$$V(c) = E[S_N(k+1) \mid S_N(k) = c], \quad |c| \leq 1. \qquad [1.1.25]$$

The RFs for evolution processes also have the representations

$$\begin{aligned} &V^{\pm}(p_{\pm}) = p_{\pm} - V_0^{\pm}(p_{\pm}), \quad 0 \leq p_{\pm} \leq 1, \quad p_+ + p_- = 1, \\ &V(c) = c - V_0(c), \quad |c| \leq 1. \end{aligned} \qquad [1.1.26]$$

Finally, *the basic representation* of the RFs, determined by the equilibria $\rho_{\pm}$ and $\rho = \rho_+ - \rho_-$, is given by the relations

$$\begin{aligned} &V^{\pm}(p_{\pm}) = \rho_{\pm} + (1 - V_0)(p_{\pm} - \rho_{\pm}), \\ &V(c) = \rho + (1 - V_0)(c - \rho). \end{aligned} \qquad [1.1.27]$$

The peculiarity of representation [1.1.27] is that the RFs depend only on the fluctuations.

REMARK 1.1.1.– As can be seen from formulas [1.1.27], *the linear drift is a function of the fluctuations*, that is, it depends on the deviation from equilibrium of the current value of the EP. In this case, the linear drift for each frequency is determined only by the fluctuations of the corresponding frequency. At each step $k > 0$, the fluctuations of the evolution process tend to be reduced.

The equilibrium state ρ, defined as the invariant value of the RF

$$V^{\pm}(\rho_{\pm}) = \rho_{\pm}, \quad V(\rho) = \rho, \qquad [1.1.28]$$

ensures the convergence, by $k \to \infty$, of the evolution processes to the corresponding equilibria.

THEOREM 1.1.1.– *With an additional limitation of the parameter V_0:*

$$0 < V_0 < 2, \qquad [1.1.29]$$

for any initial conditions

$$|C(0)| < 1, \quad 0 < \rho_\pm(0) < 1, \qquad [1.1.30]$$

the binary and frequency evolution processes, given by the difference evolution equations [1.1.17] *and* [1.1.18], *respectively, converge, by $k \to \infty$, to an equilibrium state:*

$$\lim_{k\to\infty} C(k) = \rho, \quad \lim_{k\to\infty} P_\pm(k) = \rho_\pm. \qquad [1.1.31]$$

Proof of Theorem 1.1.1.– The proof is based on the use of Lyapunov function (Krasovskii 1963) given by a quadratic function and calculated by means of the difference evolution equation [1.1.18].

The notation for the fluctuations is written as

$$\widehat{P}_\pm(k) := P_\pm(k) - \rho_\pm.$$

Then equation [1.1.18] becomes

$$\Delta\widehat{P}_\pm(k+1) = -V_0 \cdot \widehat{P}_\pm(k), \quad k \geq 0, \quad 0 < V_0 < 2. \qquad [1.1.32]$$

Now, we calculate the increments of the quadratic function of fluctuations at $(k+1)$th stage, taking into account the difference evolutional equation [1.1.32]:

$$[\widehat{P}_+(k+1)]^2 - [\widehat{P}_+(k)]^2 = -V_0(2 - V_0)[\widehat{P}_+(k)]^2, \quad k \geq 0. \qquad [1.1.33]$$

That is, the quadratic characteristic monotonically decreases when $k \to \infty$. It is also bounded below by zero and hence given by

$$\lim_{k\to\infty} P_+(k) = \rho_+.$$

The limit value is caused by the necessary condition of convergence:

$$\lim_{k\to\infty} \Delta\widehat{P}_+(k) = 0.$$

□

1.1.4. *Stochastic difference equations*

The dynamics of EPs [1.1.17] and [1.1.18] specifies the evolution of the predictable components.

DEFINITION 1.1.3.– *The stochastic component* of binary SEs is defined by the martingale differences:

$$\begin{aligned}\Delta\mu_N(k+1) &:= \Delta S_N(k+1) - E[\Delta S_N(k+1) \mid S_N(k)] \\ &= \Delta S_N(k+1) + V_0(S_N(k) - \rho), \quad k \geq 0.\end{aligned} \qquad [1.1.34]$$

DEFINITION 1.1.4.– *The stochastic component* of frequency SEs is also defined by the martingale differences:

$$\begin{aligned}\Delta\mu_N^{\pm}(k+1) &:= \Delta S_N^{\pm}(k+1) - E[\Delta S_N^{\pm}(k+1) \mid S_N^{\pm}(k)] \\ &= \Delta S_N^{\pm}(k+1) + V_0(S_N^{\pm}(k) - \rho_{\pm}), \quad k \geq 0.\end{aligned} \qquad [1.1.35]$$

The properties of the martingale differences [1.1.34] and [1.1.35] are characterized by the first two moments.

LEMMA 1.1.1.– *The martingale differences* [1.1.34] *and* [1.1.35] *are characterized by the quadratic characteristics*

$$E[[\Delta\mu_N(k+1)]^2 \mid S_N(k)] = \sigma^2(S_N(k))/N, \quad k \geq 0, \qquad [1.1.36]$$

$$E[[\Delta\mu_N^{\pm}(k+1)]^2 \mid S_N^{\pm}(k)] = \breve{\sigma}^2(S_N^{+}(k), S_N^{-}(k))/N, \quad k \geq 0. \qquad [1.1.37]$$

Here the quadratic characteristics of the martingale differences are determined by the RFs [1.1.24]–[1.1.27]*:*

$$\begin{aligned}&\breve{\sigma}^2(p_+, p_-) := V^+(p_+)V^-(p_-), \\ &\sigma^2(c) = 1 - V^2(c), \quad p_{\pm} = \frac{1}{2}[1 \pm c].\end{aligned} \qquad [1.1.38]$$

In addition, there is a connection between the quadratic characteristics

$$\sigma^2(c) = 4\breve{\sigma}^2(p_+, p_-), \quad c = p_+ - p_-. \qquad [1.1.39]$$

Proof of Lemma 1.1.1.– First of all, the quadratic characteristics [1.1.36] and [1.1.37] are calculated by the formulae

$$E[[\Delta\mu_N(k+1)]^2 \mid S_N(k)=c] \\ = D[\,S_N(k+1) \mid S_N(k)=c\,], \quad k \geq 0. \qquad [1.1.40]$$

$$E[[\Delta\mu_N^{\pm}(k+1)]^2 \mid S_N^{\pm}(k)=p_{\pm}] \\ = D[\,S_N^{\pm}(k+1) \mid S_N^{\pm}(k)=p_{\pm}\,], \quad k \geq 0. \qquad [1.1.41]$$

Next, the following obvious relations are used:

$$E[\,(\delta_n^{\pm}(k+1))^2 \mid S_N^{\pm}(k)=p_{\pm}\,] \\ = E[\,\delta_n^{\pm}(k+1) \mid S_N^{\pm}(k)=p_{\pm}\,] \\ = E[\,S_N^{\pm}(k+1) \mid S_N^{\pm}(k)=p_{\pm}\,] = V^{\pm}(p_{\pm}), \quad k \geq 0, \qquad [1.1.42]$$

and also

$$D[\,S_N^{\pm}(k+1) \mid S_N^{\pm}(k)=p_{\pm}\,] = \frac{1}{N} D[\,\delta_n^{\pm}(k+1) \mid S_N^{\pm}(k)=p_{\pm}\,], \\ k \geq 0. \qquad [1.1.43]$$

Taking into account the relations [1.1.42] and [1.1.43], we obtain

$$D[\,\delta_n^{\pm}(k+1) \mid S_N^{\pm}(k)=p_{\pm}\,] = V^{+}(p_{+})V^{-}(p_{-}), \quad k \geq 0. \qquad [1.1.44]$$

Similarly, in the binary version, the following obvious relationships are used:

$$E[\,(\delta_n(k+1))^2 \mid S_N(k)=c\,] \equiv 1, \quad 1 \leq n \leq N, \quad k \geq 0, \qquad [1.1.45]$$

$$D[\,S_N(k+1) \mid S_N(k)=c\,] = \frac{1}{N} D[\,\delta_n(k+1) \mid S_N(k)=c\,] \\ 1 \leq n \leq N, \quad k \geq 0. \qquad [1.1.46]$$

The formulae [1.1.45] and [1.1.46], as well as the relations

$$E[\,\delta_n(k+1) \mid S_N(k)=c\,] = E[\,S_N(k+1) \mid S_N(k)=c\,] = V(c), \\ 1 \leq n \leq N, \quad k \geq 0, \qquad [1.1.47]$$

$$D[\,\delta_n(k+1) \mid S_N(k)=c\,] = 1 - V^2(c), \quad k \geq 0, \qquad [1.1.48]$$

complete the proof of Lemma 1.1.1.

The relation of quadratic characteristics [1.1.39] implies from the equality

$$V^{\pm}(p_{\pm}) = \frac{1}{2}[1 \pm V(c)], \quad p_{\pm} = \frac{1}{2}[1 \pm c]. \qquad [1.1.49]$$

□

PROPOSITION 1.1.2.– (*Basic assumption 2.*) The SEs, given by averaged sums of the sampling variables [1.1.1] and [1.1.3] are determined by the solutions of the SDEs

$$\Delta S_N^{\pm}(k+1) = -V_0(S_N^{\pm}(k) - \rho_{\pm}) + \Delta\mu_N^{\pm}(k+1), \quad k \geq 0, \quad [1.1.50]$$

$$\Delta S_N(k+1) = -V_0(S_N(k) - \rho) + \Delta\mu_N(k+1), \quad k \geq 0, \quad [1.1.51]$$

where the stochastic component is determined by the martingale differences that are characterized by the first two moments

$$E\,[\Delta\mu_N^{\pm}(k+1)] \equiv 0, \quad E\,[\Delta\mu_N(k+1)] \equiv 0, \quad k \geq 0, \qquad [1.1.52]$$

$$E\,[(\Delta\mu_N^{\pm}(k+1))^2 \,|\, S_N^{\pm}(k) = p_{\pm}] = \breve{\sigma}^2(p_+, p_-)/N,$$
$$E\,[(\Delta\mu_N(k+1))^2 \,|\, S_N(k) = c] = \sigma^2(c)/N, \quad k \geq 0. \qquad [1.1.53]$$

The quadratic characteristics [1.1.53] are defined in [1.1.38].

1.1.5. *Convergence to the equilibrium state*

The steady state of the SEs is determined by the values $\rho_{\pm}$ and $\rho = \rho_+ - \rho_-$ of the RF equilibria, defined by the condition expectations [1.1.24] and [1.1.25].

Next, Kolmogorov's theorem on the strong law of large numbers is used for the sums of independent and identically distributed random variables (Shiryaev 2011, Ch. 4, section 2).

Theorem *(Kolmogorov). Let δ_n, $1 \leq n \leq N$, be a sequence of independent and identically distributed random variables with $E|\delta_1| < \infty$. Then*

$$\frac{1}{N}\sum_{n=1}^{N} \delta_n \xrightarrow{P1} E\delta_1, \quad N \to \infty. \qquad [1.1.54]$$

The use of Kolmogorov's theorem on the convergence with probability 1 of sums of independent and identically distributed random variables in the dynamic model of SEs will require an add-on analysis and the method of induction.

THEOREM 1.1.2.– *The convergence with probability 1 of the initial conditions*

$$S_N^{\pm}(0) \xrightarrow{P1} \rho_{\pm}, \quad N \to \infty, \qquad [1.1.55]$$

provides the convergence of frequency SEs with probability 1 at each finite stage $k > 0$:

$$S_N^{\pm}(k) \xrightarrow{P1} \rho_{\pm}, \quad N \to \infty \quad \forall k > 0, \qquad [1.1.56]$$

as well as the convergence of binary SEs

$$S_N(k) \xrightarrow{P1} \rho, \quad N \to \infty \quad \forall k > 0. \qquad [1.1.57]$$

Proof of Theorem 1.1.2.– The proof is carried out by induction in the stages $k \geq 0$.

First of all, Kolmogorov's theorem is applied to the stochastic components of the SDEs [1.1.50] and [1.1.51]. Using the martingale differences [1.1.34] and [1.1.35], it is obvious that they are represented by averaged sums of independent and equally distributed random variables for any fixed $k \geq 0$. In addition, the zero value of the first moments [1.1.52] means that according to Kolmogorov's theorem, there is a convergence with probability 1 at each finite stage $k \geq 1$:

$$\Delta\mu_N^{\pm}(k+1) \xrightarrow{P1} 0, \quad N \to \infty, \qquad [1.1.58]$$

$$\Delta\mu_N(k+1) \xrightarrow{P1} 0, \quad N \to \infty. \qquad [1.1.59]$$

Under the condition of convergence, with probability 1, of SEs [1.1.56] and [1.1.57] for a fixed $k > 0$, the SDE solution [1.1.50] is used to represent the SE in the next step $k + 1$:

$$S_N^{\pm}(k+1) = S_N^{\pm}(k) - V_0^{\pm}(S_N^{\pm}(k)) + \Delta\mu_N^{\pm}(k+1), \quad k > 0. \qquad [1.1.60]$$

The state of SEs at the $(k + 1)$th stage is determined by the RF:

$$S_N^{\pm}(k+1) = V^{\pm}(S_N^{\pm}(k)) + \Delta\mu_N^{\pm}(k+1), \quad k \geq 0. \qquad [1.1.61]$$

Taking into account [1.1.27], we have

$$S_N^{\pm}(k+1) = \rho_{\pm} + (1 - V_0)(S_N^{\pm}(k) - \rho_{\pm}) + \Delta\mu_N^{\pm}(k+1), \; k \geq 0. \qquad [1.1.62]$$

Using the assumption of induction, $S_N^{\pm}(k) - \rho_{\pm} \xrightarrow{P1} 0$ by $N \to \infty$.

The convergence to zero with probability 1 of the martingale differences [1.1.58] ensures the convergence with probability 1 of SEs at the next, $(k+1)$th stage:

$$S_N^{\pm}(k+1) \xrightarrow{P1} \rho_{\pm}, \quad \text{by} \quad N \to \infty. \qquad [1.1.63]$$

Theorem 1.1.2 is proved. □

1.1.6. *Normal approximation of the stochastic component*

The asymptotic property of the martingale differences [1.1.34] and [1.1.35] establishes the following normal approximation.

THEOREM 1.1.3.– *In the conditions of Theorem 1.1.2, a convergence in distribution takes place:*

$$\sqrt{N}\,\Delta\mu_N^{\pm}(k+1) \xrightarrow{d} \breve{\sigma} W^{\pm}(k+1), \quad N \to \infty, \qquad [1.1.64]$$

$$\sqrt{N}\,\Delta\mu_N(k+1) \xrightarrow{d} \sigma W(k+1), \quad N \to \infty, \qquad [1.1.65]$$

where

$$\sigma^2 = 1 - \rho^2, \quad \breve{\sigma}^2 = \rho_+\rho_-, \quad \sigma^2 = 4\breve{\sigma}^2, \qquad [1.1.66]$$

and $W(k)^{\pm}$ and $W(k)$ are normally distributed standard random variables with $EW^{\pm}(k) = 0$, $E(W^{\pm}(k))^2 = 1$, for all $k \geq 1$.

Proof of Theorem 1.1.3.– Consider the normalized frequency fluctuations

$$\zeta_N^{\pm}(k) := \sqrt{N}(S_N^{\pm}(k) - \rho_{\pm}), \qquad [1.1.67]$$

$$\zeta_N(k) := \sqrt{N}(S_N(k) - \rho). \qquad [1.1.68]$$

Then, in accordance with Proposition 1.1.2, the normalized SEs [1.1.68] are given by the normalized martingale differences

$$\begin{aligned} \sqrt{N}\Delta\mu_N^{\pm}(k+1) &= \zeta_N^{\pm}(k+1) - (1 - V_0)\zeta_N^{\pm}(k), \\ \sqrt{N}\Delta\mu_N(k+1) &= \zeta_N(k+1) - (1 - V_0)\zeta_N(k), \end{aligned} \qquad [1.1.69]$$

the quadratic characteristics of which are given by the relations

$$E\left[(\sqrt{N}\Delta\mu_N^{\pm}(k+1))^2 \mid S_N(k)\right] = \breve{\sigma}_N^2(S_N^+(k), S_N^-(k)), \qquad [1.1.70]$$

$$\breve{\sigma}^2(p_+, p_-) := V^+(p_+)V^-(p_-),$$

$$V^{\pm}(p_\pm) = [\rho_\pm + (1 - V_0)(p_\pm - \rho_\pm)],$$

$$E\left[(\sqrt{N}\Delta\mu_N(k+1))^2 \mid S_N(k)\right] = \sigma^2(S_N(k)),$$

$$\sigma^2(c) := 1 - V^2(c), \quad V(c) := \rho + (1 - V_0)(c - \rho). \qquad [1.1.71]$$

The passage to limit, by $N \to \infty$, in relations [1.1.70] and [1.1.71], gives the following asymptotic representations (see Theorem 1.1.2):

$$\breve{\sigma}^2(S_N^+(k), S_N^-(k)) \xrightarrow{P1} \breve{\sigma}^2, \quad N \to \infty, \qquad [1.1.72]$$

and also

$$\sigma^2(S_N(k)) \xrightarrow{P1} \sigma^2, \quad N \to \infty. \qquad [1.1.73]$$

uniformly by $k \leq K < \infty$.

The normalized martingale differences [1.1.69] are also determined by the sum of independent and equally distributed random variables with mathematical expectations zero and quadratic characteristics given by the relations [1.1.36] and [1.1.37], equally distributed by a fixed $k \geq 1$. The values limitation of the random variables that make up the martingale difference [1.1.69] ensures the fulfillment of the Lindeberg condition necessary (Shiryaev 2011, Ch. 3, section 4.2) in order to use the central limit theorem.

The convergence of quadratic characteristics [1.1.72] and [1.1.73] substantiates the assertion of Theorem 1.1.3. □

REMARK 1.1.2.– The condition of uniform infiniteness, by probability, of the summands (Borovskikh and Korolyuk, 1997, section 5.2)

$$P\{\max_{1\leq n\leq N} |\beta_n(k)| > \varepsilon\sqrt{N}\} \to 0, \quad N \to \infty \quad \forall \varepsilon > 0, \qquad [1.1.74]$$

is equivalent to the classical Lindeberg condition.

The limit relations [1.1.64] and [1.1.65] of Theorem 1.1.3 give a reason to use a stochastic approximation for the SE processes [1.1.1]–[1.1.3] in the following form (Koroliouk 2015a,b).

PROPOSITION 1.1.3.– The processes of normal autoregression, approximating the SE sequences [1.1.1]–[1.1.3], are given by the relations

$$\Delta\widehat{S}_N^{\pm}(k+1) = -V_0\widehat{S}_N^{\pm}(k) + \frac{\breve{\sigma}}{\sqrt{N}}W(k+1), \quad \breve{\sigma}^2 = \rho_+\rho_-, \qquad [1.1.75]$$

$$\Delta\widehat{S}_N(k+1) = -V_0\widehat{S}_N(k) + \frac{\sigma}{\sqrt{N}}W(k+1), \quad \sigma^2 = 1-\rho^2. \qquad [1.1.76]$$

The stochastic component $W(k)$, $k \geq 1$, is represented by jointly independent, normally distributed standard random variables.

REMARK 1.1.3.– For normalized SE fluctuations, under the Lindeberg condition [1.1.74], the corresponding processes of normal autoregression are given by the solutions of the SDEs

$$\Delta\zeta_N^{\pm}(k+1) = -V_0\zeta_N^{\pm}(k) + \breve{\sigma}\Delta W(k+1), \qquad [1.1.77]$$

$$\Delta\zeta_N(k+1) = -V_0\zeta_N(k) + \sigma\Delta W(k+1). \qquad [1.1.78]$$

1.2. Binary SEs with nonlinear regression

Nonlinear RFs contain the linear factor of the fluctuation and preserve the convergence of SEs to the equilibrium state and the normal approximation of the stochastic component as the sample volume $N \to \infty$ (sections 1.2.4 and 1.2.5).

1.2.1. *Basic assumptions*

As in section 1.1, the SEs are determined by averaging sums of *sampling random variables*.

The binary SEs with nonlinear regression are determined by averaging sums of sampling random variables:

$$S_N(k) := \frac{1}{N}\sum_{n=1}^{N}\delta_n(k), \quad k \geq 0, \qquad [1.2.1]$$

which assume the binary values ± 1.

The frequency SEs are given by the averaged sums of indicators

$$S_N^{\pm}(k) := \frac{1}{N}\sum_{n=1}^{N}\delta_n^{\pm}(k), \quad k \geq 0. \qquad [1.2.2]$$

The sampling random variables $\delta_n^{\pm}(k)$, $0 \leq n \leq N$, $k \geq 0$, can take two values ± 1:

$$\delta_n^{\pm}(k) := I\{\delta_n(k) = \pm 1\}, \quad 0 \leq n \leq N, \quad k \geq 0. \qquad [1.2.3]$$

The obvious frequency representation of binary SEs [1.2.1] is

$$S_N(k) = S_N^{+}(k) - S_N^{-}(k), \ k \geq 0. \qquad [1.2.4]$$

The evolution, by $k \geq 0$, of SEs [1.2.2] and [1.2.4], is set by the EPs:

$$P_{\pm}(k+1) := E[S_N^{\pm}(k+1) \mid S_N^{\pm}(k) = P_{\pm}(k)], \quad k \geq 0, \qquad [1.2.5]$$

$$C(k+1) := E[S_N(k+1) \mid S_N(k) = C(k)], \quad k \geq 0. \qquad [1.2.6]$$

Obviously, EPs [1.2.5] and [1.2.6] do not depend on the sample size N and determine the predicted components.

The balance condition

$$P_{+}(k) + P_{-}(k) \equiv 1 \quad \forall k \geq 0,$$

and the frequency representation of the binary EPs

$$P_{+}(k) - P_{-}(k) = C(k) \quad \forall k \geq 0, \qquad [1.2.7]$$

allow us to obtain the representation of the EPs frequency by using binary EPs:

$$P_{\pm}(k) = \frac{1}{2}[1 \pm C(k)], \quad k \geq 0. \qquad [1.2.8]$$

The dynamics of the EPs [1.2.5] and [1.2.6] is postulated by the choice of RFIs, which is given by the DEEs [1.2.11] and [1.2.14].

Taking into account the predicted nature of the EPs [1.2.5] and [1.2.6], the RFIs are also determined by *conditional mathematical expectations*

$$E[\Delta S_N(k+1) \mid S_N(k) = c] = -\, V_0(c)(c - \rho),$$
$$|c| \leq 1, \quad |\rho| < 1, \qquad [1.2.9]$$

$$E[\Delta S_N^{\pm}(k+1) \mid S_N^{\pm}(k) = p_{\pm}] = -\, V_0(p_{\pm})(p_{\pm} - \rho_{\pm}),$$
$$0 \leq p_{\pm} \leq 1, \quad p_{+} + p_{-} = 1. \qquad [1.2.10]$$

The nonlinear factors are determined by the relationships [1.2.12] and [1.2.14].

Since the drift parameters $\rho_\pm$, $\rho = \rho_+ - \rho_-$, are determined using the equilibrium state of the regression of increments, the increments of evolution processes are zero at the equilibrium points $\rho_\pm$ and ρ.

PROPOSITION 1.2.1.– (*Basic assumption 1*). The frequency EPs [1.2.5] are determined by the solutions of the DEEs

$$\Delta P_\pm(k+1) = -V_0(P(k))[P_\pm(k) - \rho_\pm], \quad k \geq 0. \tag{1.2.11}$$

The RFI is given with a nonlinear factor

$$V_0(p) := V_0 \cdot p_+ p_-, \quad 0 \leq p_\pm \leq 1, \quad p_+ + p_- = 1, \quad V_0 > 0. \tag{1.2.12}$$

The binary EPs [1.2.5] are determined by the solutions of the DEEs

$$\Delta C(k+1) = -V_0(C(k))(C(k) - \rho), \quad k \geq 0. \tag{1.2.13}$$

The binary increment RF is given with a nonlinear multiplier

$$V_0(c) := V_0(1 - c^2), \quad |c| \leq 1, \quad V_0 > 0. \tag{1.2.14}$$

REMARK 1.2.1.– The fundamental principle of the dynamics of EPs—"stimulation-deterrence"—also remains relevant for the DEEs [1.2.11] and [1.2.13] with the corresponding nonlinear RFIs [1.2.12] and [1.2.14].

The linear multiplier [1.2.10] has a presentation

$$p_+ - \rho_+ = -(p_- - \rho_-) = \rho_- p_+ - \rho_+ p_-. \tag{1.2.15}$$

So, the DEEs [1.2.11] and [1.2.13] have RFIs, which are determined by the fluctuations (relative to the corresponding equilibria) of binary and frequency EPs.

1.2.2. *Equilibrium*

Taking into account the dynamics of the EPs, which is determined by the DEEs [1.2.11] and [1.2.12], the EPs are also determined *recursively* by the RFs:

$$C(k+1) = C(k) - V_0(C(k))(C(k) - \rho), \quad k \geq 0, \tag{1.2.16}$$

$$P_\pm(k+1) = P_\pm(k) - V_0(P(k))(P_\pm(k) - \rho_\pm), \quad k \geq 0. \tag{1.2.17}$$

PROPOSITION 1.2.2.– (*Basic assumption 2*). The EPs [1.2.5] and [1.2.6] are given by the recursive formulae

$$C(k+1) - \rho = V_\sigma(C(k))(C(k) - \rho), \quad k \geq 0, \qquad [1.2.18]$$

$$P_\pm(k+1) - \rho_\pm = V_\sigma(P_\pm(k))(P_\pm(k) - \rho_\pm), \quad k \geq 0. \qquad [1.2.19]$$

The nonlinear factors are given by the relations

$$V_\sigma(c) := 1 - V_0(c) = 1 - V_0(1 - c^2), \qquad [1.2.20]$$

$$V_\sigma(p_\pm) = 1 - 4V_0 \cdot p_\pm(1 - p_\pm), \quad p_\pm = \frac{1}{2}[1 \pm c]. \qquad [1.2.21]$$

The presence of the equilibrium state of the RF ensures the EP convergence to the equilibrium state, as $k \to \infty$, with the additional condition $0 < V_0 < 2$.

THEOREM 1.2.1.– *For any initial conditions* $|C(0)| < 1$, *the binary EPs converge to the equilibrium state*

$$\lim_{k\to\infty} C(k) = \rho. \qquad [1.2.22]$$

Proof of theorem 1.2.1.– The proof is based on the use of a Lyapunov function, as in the proof of Theorem 1.1.1.

Instead of the equality [1.1.33], we have a relation

$$[\widehat{C}(k+1)]^2 - [\widehat{C}(k)]^2 = -V_0(c)(2 - V_0(c))[\widehat{C}(k)]^2. \qquad [1.2.23]$$

Here (see [1.2.14]),

$$V_0(c) := V_0(1 - c^2) \leq V_0 < 2.$$

That is, the Lyapunov quadratic function $[\widehat{C}(k)]^2$ monotonously decreases by $k \to \infty$. It is also limited, from below, by zero. So, there is a limit

$$\lim_{k\to\infty} [C(k) - \rho] = 0. \qquad [1.2.24]$$

The value zero of the limit [1.2.24] justifies the necessary condition of convergence

$$\lim_{k\to\infty} \Delta C(k+1) = 0, \qquad [1.2.25]$$

which is provided by the equilibrium state ρ. □

1.2.3. *Stochastic difference equations*

The dynamics of EPs [1.2.5] and [1.2.6] is determined by the DEEs [1.2.11]–[1.2.13].

DEFINITION 1.2.1.– *The stochastic component* of the SEs *determined by the martingale differences* is

$$\begin{aligned}\Delta\mu_N^{\pm}(k+1) &:= \Delta S_N^{\pm}(k+1)\\ &\quad - E[\Delta S_N^{\pm}(k+1) \mid S_N^{\pm}(k)], \quad k \geq 0,\end{aligned} \qquad [1.2.26]$$

for frequency SE $S_N^{\pm}(k)$, $k \geq 0$, and

$$\begin{aligned}\Delta\mu_N(k+1) &:= \Delta S_N(k+1)\\ &\quad - E[\Delta S_N(k+1) \mid S_N(k)], \quad k \geq 0,\end{aligned} \qquad [1.2.27]$$

for the binary SE $S_N(k)$, $k \geq 0$.

Taking into account the DEEs [1.2.11] and [1.2.13], the martingale differences [1.2.26] and [1.2.27] have the following representations:

$$\begin{aligned}\Delta\mu_N^{\pm}(k+1) &= \Delta S_N^{\pm}(k+1)\\ &\quad + V_0^{\pm}(S_N^{\pm}(k))(S_N^{\pm}(k) - \rho_{\pm}), \ k \geq 0,\end{aligned} \qquad [1.2.28]$$

and

$$\begin{aligned}\Delta\mu_N(k+1) &:= \Delta S_N(k+1)\\ &\quad + V_0^{\pm}(S_N(k))(S_N(k) - \rho), \ k \geq 0.\end{aligned} \qquad [1.2.29]$$

The nonlinear RFI factors have representations [1.2.12] and [1.2.14].

The properties of the martingale differences [1.2.28] and [1.2.29] are characterized by the first two moments.

LEMMA 1.2.1.– *The martingale differences* [1.2.28] *and* [1.2.29] *are determined by quadratic characteristics*

$$E[[\Delta\mu_N^{\pm}(k+1)]^2 \mid S_N(k)] = \breve{\sigma}^2(S_N^{+}(k), S_N^{-}(k))/N, \quad k \geq 0, \qquad [1.2.30]$$

$$E[[\Delta\mu_N(k+1)]^2 \mid S_N(k)] = \sigma^2(S_N(k))/N, \quad k \geq 0. \qquad [1.2.31]$$

The quadratic characteristics are given by the relations

$$\breve{\sigma}^2(p_+, p_-) := V^+(p_+)V^-(p_-),$$
$$\sigma^2(c) := 1 - V^2(c). \qquad [1.2.32]$$

Proof of Lemma 1.2.1.– The proof is conducted in a similar way to the proof of Lemma 1.1.1, taking into account the nonlinear RFs in the representation of the martingale differences [1.2.30] and [1.2.31]. □

PROPOSITION 1.2.3.– (*Basic assumption 3.*) The SEs, given by the averaged sums of the sample values [1.2.1] and [1.2.2], are determined by the solutions of the SDEs:

$$\Delta S_N^{\pm}(k+1) = -V_0(S_N^{\pm}(k))[S_N^{\pm}(k) - \rho_{\pm}] + \Delta\mu_N^{\pm}(k+1), \quad k \geq 0, \qquad [1.2.33]$$

$$\Delta S_N(k+1) = -V_0(S_N(k))[S_N(k) - \rho] + \Delta\mu_N(k+1), \quad k \geq 0. \qquad [1.2.34]$$

The stochastic components of the SDEs [1.2.33] and [1.2.34] are characterized by quadratic characteristics [1.2.30] and [1.2.31] and zero first moments

$$E\Delta\mu_N(k+1) = E\Delta\mu_N^{\pm}(k+1) \equiv 0 \quad \forall k \geq 0. \qquad [1.2.35]$$

1.2.4. *Convergence to the equilibrium state*

As stated before, the balanced state of SEs is determined by the equilibria $\rho_{\pm}$, $\rho = \rho_+ - \rho_-$ $(\rho_+ + \rho_- = 1)$, of the RFs, given by the conditional mathematical expectations (see [1.2.6] and [1.2.7]):

$$V(c) := E[S_N(k+1) \mid S_N(k) = c] = \rho + V_\sigma(c)(c - \rho), \ k \geq 0, \qquad [1.2.36]$$

$$V^{\pm}(p_{\pm}) := E[S_N^{\pm}(k+1) \mid S_N^{\pm}(k) = p_{\pm}]$$
$$= \rho_{\pm} + V_\sigma(p_{\pm})(p_{\pm} - \rho_{\pm}), \quad k \geq 0. \qquad [1.2.37]$$

The nonlinear factor $V_\sigma(c)$ is determined by the relations

$$V_\sigma(c) := 1 - V_0(c),$$
$$V_0(c) := V_0(1 - c^2), \quad V_0(p_{\pm}) := 4V_0 \cdot p_{\pm}(1 - p_{\pm}). \qquad [1.2.38]$$

Using Kolmogorov's theorem (see [1.1.63]) on the convergence with probability 1 of averaged sums of sampling variables, which are set recursively with RFs [1.2.36] and [1.2.37],

$$S_N(k+1) = \rho + V_\sigma(S_N(k))(S_N(k) - \rho) + \Delta\mu_N(k+1), \quad k \geq 0,$$
$$S_N^{\pm}(k+1) = \rho_{\pm} + V_\sigma(S_N^{\pm}(k))(S_N^{\pm}(k) - \rho_{\pm}) + \Delta\mu_N^{\pm}(k+1), \quad [1.2.39]$$

requires additional analysis taking into account the nonlinearity of the factors [1.2.39] and the induction method application by $k \geq 0$.

THEOREM 1.2.2.– *The convergence with probability 1 of initial conditions*

$$S_N^{\pm}(0) \xrightarrow{P1} \rho_{\pm}, \quad N \to \infty, \qquad [1.2.40]$$

provides the convergence with probability 1 of SEs at each finite stage $k > 0$:

$$S_N(k) \xrightarrow{P1} \rho, \quad N \to \infty \quad \forall k > 0. \qquad [1.2.41]$$

Proof of Theorem 1.2.2.– The proofs of Theorems 1.1.2 and 1.2.2 are based on the induction, by $k \geq 0$, taking into account the representation of the martingale differences [1.2.32] by averaged sums of sampling variables:

$$\Delta\mu_N(k+1) := \frac{1}{N}\sum_{n=1}^{N} \beta_n(k+1), \quad k \geq 0, \qquad [1.2.42]$$

which take two possible values

$$\beta_n(k+1) = \pm 1 - V(c) \text{ with probabilities } P_{\pm}(k+1)$$
$$= \frac{1}{2}[1 \pm V(C(k))], \quad k \geq 0.$$

Obviously,

$$E\beta_n(k+1) \equiv 0, \ E[[\beta_n(k+1)]^2 \mid S_N(k) = c] = 1 - V^2(c) \quad \forall k \geq 0. \qquad [1.2.43]$$

Consequently, by Kolmogorov's theorem, there is a convergence with probability 1 at each finite stage $k \geq 0$:

$$\Delta\mu_N(k+1) \xrightarrow{P1} 0 \ \text{ by } \ N \to \infty. \qquad [1.2.44]$$

Now, using the assumption of induction, there is a convergence with probability 1 of the predictable component:

$$V_\sigma(S_N(k))(S_N(k) - \rho) \xrightarrow{P1} 0 \quad \text{by} \quad N \to \infty. \qquad [1.2.45]$$

And together with the convergence [1.2.45] of the martingale differences, there is a convergence with probability 1:

$$S_N(k) - \rho \xrightarrow{P1} 0 \quad \text{by} \quad N \to \infty. \qquad [1.2.46]$$

Theorem 1.2.2 is proved. □

1.2.5. *Normal approximation of the stochastic component*

The presence of the equilibrium state, formulated in Theorem 1.2.2, is an important, but also the simplest property of SEs. The simplified description of SE trajectories can give an asymptotic analysis of the stochastic component of SEs by increasing the sample size $N \to \infty$.

The asymptotic property of the martingale differences [1.2.31] and [1.2.32] establishes the following.

THEOREM 1.2.3.– *Under the conditions of Theorem 1.2.2, we obtain the following convergences in distribution:*

$$\sqrt{N}\Delta\mu_N(k+1) \xrightarrow{d} \sigma W(k+1), \quad N \to \infty, \qquad [1.2.47]$$

$$\sqrt{N}\Delta\mu_N^{\pm}(k+1) \xrightarrow{d} \breve{\sigma} W(k+1), \quad N \to \infty. \qquad [1.2.48]$$

Here the quadratic characteristics σ^2 and $\breve{\sigma}^2$ are determined by the relations

$$\sigma^2 = 1 - \rho^2, \quad \breve{\sigma}^2 = \rho_+\rho_-, \quad \sigma^2 = 4\breve{\sigma}^2. \qquad [1.2.49]$$

and $W(k)$ are the standard normally distributed random variables

$$EW(k) \equiv 0, \quad EW^2(k) \equiv 1 \quad \forall k \geq 1. \qquad [1.2.50]$$

Proof of Theorem 1.2.3.– This is similar to the proof of Theorem 1.1.3, taking into account the asymptotic behavior of quadratic characteristics of the normalized martingale differences:

$$E[(\sqrt{N}\Delta\mu_N(k+1))^2 \mid S_N(k)] = \sigma^2(S_N(k)), \qquad [1.2.51]$$

The quadratic characteristic of the normalized martingale differences is given by

$$\sigma^2(c) := 1 - V^2(c), \quad V(c) := \rho + V_\sigma(c)(c - \rho), \qquad [1.2.52]$$

where $V_\sigma(c)$ is defined by formula [1.1.27].

By definition of equilibrium, we have $V(\rho) = \rho$, and hence the passage to limit by $N \to \infty$ gives the asymptotic representation of the quadratic characteristic

$$\lim_{N\to\infty} \sigma^2(S_N(k)) = \sigma^2(\rho) = 1 - \rho^2 = \sigma^2, \qquad [1.2.53]$$

uniformly by $k \leq K \leq \infty$.

The recursive representation of the SE [1.2.39] is used to specify the normalized martingale differences [1.2.51] using the normalized fluctuations of the SEs:

$$\zeta_N(k) := \sqrt{N}[S_N(k) - \rho], \quad k \geq 0. \qquad [1.2.54]$$

The normalized martingale differences have the representation

$$\sqrt{N}\Delta\mu_N(k+1) = \zeta_N(k+1) - V_\sigma(S_N(k))\zeta_N(k), \quad k \geq 0. \quad [1.2.55]$$

Consequently, the martingale differences are also given by normalized sums of random sampling variables, equally distributed by fixed $k \geq 0$ with zero mathematical expectation. The boundedness of the values of random variables, presented in the normalized martingale differences [1.2.55], ensures the validity of the Lindeberg condition, necessary for the application of the central limit theorem (Shiryaev 2011). The convergence of quadratic characteristics [1.2.53] provides the assertion of Theorem 1.2.3. □

The asymptotic representation of the normalized martingale differences [1.2.47] and [1.2.48] provides the basis for using the stochastic approximation of SEs in the following form.

PROPOSITION 1.2.4.– SEs are determined by the solution of the SDEs:

$$\Delta S_N(k+1) = - V_0(S_N(k))[S_N(k) - \rho] + \left(\sigma\sqrt{N}\right)W(k+1), \; k \geq 0, \qquad [1.2.56]$$

or

$$\Delta S_N^{\pm}(k+1) = -V_0(S_N^{\pm}(k))[S_N^{\pm}(k) - \rho_{\pm}] \\ + (\breve{\sigma}\sqrt{N})W(k+1),\ k \geq 0. \quad [1.2.57]$$

PROPOSITION 1.2.5.– The normalized fluctuations [1.2.54] are given by the solution of linear SDEs

$$\Delta\zeta_N(k+1) = -V_0\sigma^2\zeta_N(k) + \sigma W(k+1), \quad k \geq 0, \quad [1.2.58]$$

and

$$\Delta\zeta_N^{\pm}(k+1) = -V_0\breve{\sigma}^2\zeta_N^{\pm}(k) + \breve{\sigma}W(k+1), \quad k \geq 0. \quad [1.2.59]$$

1.3. Multivariate statistical experiments

The study of SEs, generated by binary sampling variables that take two values, plays a significant role in their analysis and applications with interpretations in real decision-making models.

However, the number of elementary decisions that defines SEs may be more than 2 and be determined by a finite number $M+1$, where $M \geq 1$.

Multivariate statistical experiments (MSEs) are studied in the assumption that a finite number of possible values is determined by a state space of values $E := \{e_0, e_1, \dots, e_M\}$, $M \geq 1$, where $e_0 = (1, 0, \dots, 0), \dots, e_M = (0, 0, ..., 1)$.

1.3.1. *Regression function of increments*

MSEs are generated by averaged sums of sampling random variables $\delta_n(k)$, $1 \leq n \leq N$, with a finite number of values $M+1$ ($M \geq 1$) from a set $E = \{e_0, e_1, \dots, e_M\}$:

$$S_N(k+1) := \frac{1}{N}\sum_{n=1}^{N} \delta_n(k), \quad k \geq 0. \quad [1.3.1]$$

Let us introduce the binary random variables that take two values:

$$\delta_n^{(m)}(k) := I\{\delta_n(k) = e_m\}, \quad 0 \leq m \leq M, \quad 1 \leq n \leq N. \quad [1.3.2]$$

The frequency MSEs are given by averaged sums

$$S_N^{(m)}(k) := \frac{1}{N} \sum_{n=1}^{N} \delta_n^{(m)}(k), \quad 0 \leq m \leq M, \quad k \geq 0. \qquad [1.3.3]$$

Further more, we used the following vector representation of frequency MSEs:

$$S_N(k) := \left(S_N^{(m)}(k),\, 0 \leq m \leq M \right), \quad k \geq 0, \qquad [1.3.4]$$

as well as a similar vector representation of multivariate random variables:

$$\delta_n(k) := \left(\delta_n^{(m)}(k),\, 0 \leq m \leq M \right), \quad 1 \leq n \leq N, \quad k \geq 0. \qquad [1.3.5]$$

The dynamics of MSEs by $k \geq 0$ are first considered for EPs, determined by the conditional mathematical expectations:

$$P_m(k+1) := E[S_N^{(m)}(k+1) \mid S_N(k) = P(k)],$$
$$0 \leq m \leq M, \quad k \geq 0. \ [1.3.6]$$

The dynamics of EP [1.3.6] is determined for increments

$$\Delta P_m(k+1) := P_m(k+1) - P_m(k), \quad 0 \leq m \leq M, \quad k \geq 0.$$

PROPOSITION 1.3.1.– (*Basic assumption 1*). The frequency multivariate linear EP [1.3.6] is determined by the solution of the following DEE with linear RFIs:

$$\Delta P_m(k+1) = -V_0[P_m(k) - \rho_m], \quad 0 \leq m \leq M, \quad k \geq 0, \quad V_0 > 0. \qquad [1.3.7]$$

The RFIs generate EPs using the equilibria (Koroliouk and Koroliuk 2019):

$$\left(\rho_m, \quad 0 \leq m \leq M \right), \quad 0 < \rho_m < 1, \quad \sum_{m=0}^{M} \rho_m = 1.$$

The drift parameter, by definition, is positive: $V_0 > 0$.

Next, we obtain the global balance condition of the RFIs:

$$\sum_{m=0}^{M} [P_m(k) - \rho_m] \equiv 0 \quad \forall k \geq 0.$$

The nonlinear RFIs are generated in the analysis of models of SEs with Wright–Fisher normalization (Koroliouk *et al.*, 2014) in population genetics (see section 1.4).

Let us denote $p := (p_m, 0 \leq m \leq M)$; $P(k) := (P_m(k), 0 \leq m \leq M)$, $k \geq 0$.

PROPOSITION 1.3.2.– (*Basic assumption* 2). The frequency multivariate nonlinear EP [1.3.6] is given by the solution of the following DEE:

$$\Delta P_m(k+1) = V_0^{(m)}(P(k)), \quad 0 \leq m \leq M, \quad k \geq 0. \qquad [1.3.8]$$

with a nonlinear RFI *(canonical representation)*

$$V_0^{(m)}(p) := V_0 \pi \left[p_m \sum_{n=0}^{M} p_n^2/\rho_n - p_m^2/\rho_m \right], \quad 0 \leq m \leq M. \qquad [1.3.9]$$

Here, the parameters

$$V_0 > 0, \quad \pi := \prod_{m=0}^{M} \rho_m, \quad \sum_{m=0}^{M} \rho_m = 1.$$

Then we obtain a global balance condition, which ensures the correctness of the DEE [1.3.8]:

$$\sum_{m=0}^{M} V_0^{(m)}(p) \equiv 0 \quad \forall p : \sum_{m=0}^{M} p_m = 1.$$

Using equilibrium, as well as the local balance condition, we can reformulate the Basic assumption 2 in the following equivalent form.

PROPOSITION 1.3.3.– (*Basic assumption 3*). The frequency multivariate nonlinear EP [1.3.6] is determined by the DEE [1.3.9] solutions with nonlinear RFIs *(fluctuation representation)*:

$$V_0^{(m)}(\widehat{p}) = V_0 p_m \sum_{n=0}^{M} p_n(\pi_n \widehat{p}_n - \pi_m \widehat{p}_m), 0 \leq m \leq M, \qquad [1.3.10]$$

with normalizing factors $\pi_m := \pi/\rho_m$, where $0 \leq m \leq M$.

Here, we use the usual symbol for the fluctuations

$$\widehat{p}_n := p_n - \rho_n, \quad 1 \leq n \leq N.$$

1.3.2. *The equilibrium state of multivariate EPs*

The representation [1.3.10] generates an equilibrium (local balance condition)

$$V_0^{(m)}(0) \equiv 0, \quad p_m = \rho_m, \quad 0 \leq m \leq M. \qquad [1.3.11]$$

The presence of the equilibrium state ensures the EP convergence to equilibrium.

THEOREM 1.3.1.– *By the condition on the drift parameter* V_0*:* $0 < V_0 < 2$, *for any initial data:* $0 < P_m(0) < 1$, $0 \leq m \leq M$, *the EPs* $P_m(k)$, $0 \leq m \leq M$, $k \geq 0$, *determined by the solutions of the DEEs* [1.3.8]–[1.3.10], *converge as* $k \to \infty$, *to the equilibria*

$$\lim_{k\to\infty} P_m(k) = \rho_m, \quad 0 \leq m \leq M. \qquad [1.3.12]$$

Proof of Theorem 1.3.1.– The Lyapunov function (Krasovskii 1963) is used, which is given by a quadratic function and is calculated on the paths of the DEE solution [1.3.7], taking into account the transition to the centered fluctuations $\widehat{P}_m(k) := P_m(k) - \rho_m$:

$$\Delta P_m(k+1) = \Delta \widehat{P}_m(k+1), \quad 0 \leq m \leq M, \quad k \geq 0. \qquad [1.3.13]$$

Its quadratic characteristic on $(k+1)$th stage has the following form:

$$[\widehat{P}_m(k+1)]^2 - [\widehat{P}_m(k)]^2 = -V_0(2 - V_0)[\widehat{P}_m(k)]^2. \qquad [1.3.14]$$

Consequently, the quadratic characteristic monotonically decreases by $k \to \infty$. The boundedness from below by zero ensures the convergence

$$\lim_{k\to\infty} \widehat{P}_m(k) = 0, \qquad [1.3.15]$$

or otherwise, the assertion [1.3.12] of Theorem 1.3.1 is fulfilled. The limit value [1.3.15] is justified by the necessary condition of the convergence:

$$\lim_{k\to\infty} \Delta \widehat{P}_m(k+1) = 0.$$

□

The question of EP convergence to the equilibrium state, determined by equilibrium $\rho = (\rho_m,\ 0 \leq m \leq M)$ with a nonlinear RFI [1.3.9], needs an additional analysis.

The global balance condition, ensuring the correctness of the DEE [1.3.8], means that m_* exist, for which $V_0^{(m_*)}(p) < 0$, and hence $\Delta P_{m_*}(k+1) < 0$, that is, the increment decreases. The limit [1.3.15] is provided by the necessary condition of the convergence $\lim_{k\to\infty} \Delta P_m(k+1) = 0$.

1.3.3. *Stochastic difference equations*

The dynamics of EPs, given by conditional mathematical expectations [1.3.6], is determined by the DEE [1.3.7] with a linear or nonlinear RFI [1.3.10].

The EP defines the predicted components of the SE.

DEFINITION 1.3.1.– The stochastic components of SEs are determined by the following martingale differences:

$$\Delta\mu_N^{(m)}(k+1) := \Delta S_N^{(m)}(k+1) - E[\Delta S_N^{(m)}(k+1) \mid S_N(k)]. \qquad [1.3.16]$$

Given the RFIs [1.3.9] and [1.3.10], the martingale differences are given by the difference between the two components:

$$\Delta\mu_N^{(m)}(k+1) = \Delta S_N^{(m)}(k+1) - V_0^{(m)}(S_N^{(m)}(k)),$$
$$0 \leq m \leq M, \quad k \geq 0. \qquad [1.3.17]$$

The predicted component is determined by the RFI [1.3.10].

The martingale differences [1.3.17] also have the following representation:

$$\Delta\mu_N^{(m)}(k+1) = S_N^{(m)}(k+1) - V^{(m)}(S_N^{(m)}(k)), \quad k \geq 0, \qquad [1.3.18]$$

with the RF

$$V^{(m)}(p) := p_m + V_0^{(m)}(p), \quad 0 \leq m \leq M. \qquad [1.3.19]$$

The presence of the equilibrium state in RFIs indicates the existence of equilibrium for the RF:

$$V^{(m)}(\rho) = \rho_m, \quad 0 \leq m \leq M.$$

The martingale differences [1.3.18] are characterized by the first two moments.

LEMMA 1.3.1.– *The quadratic characteristics of the martingale differences* [1.3.18] *are determined by a conditional mathematical expectation*

$$E\left[(\Delta\mu_N^{(m)}(k+1))^2 \mid S_N(k)\right] = \sigma_m^2(S_N(k))/N,$$

$$0 \leq m \leq M, k \geq 0. \quad [1.3.20]$$

They are expressed as

$$\sigma_m^2(p) = V^{(m)}(p)[1 - V^{(m)}(p)], \quad 0 \leq m \leq M. \quad [1.3.21]$$

The equilibrium state provides the following representation:

$$\sigma_m^2 := \sigma_m^2(\rho) = \rho_m(1 - \rho_m), \quad 0 \leq m \leq M. \quad [1.3.22]$$

Proof of Lemma 1.3.1.– The representation of quadratic characteristics [1.3.20] is used by means of the conditional dispersions of MSEs:

$$\sigma_m^2(S_N^{(m)}(k)) = D[S_N^{(m)}(k+1) \mid S_N(k)]. \quad [1.3.23]$$

The obvious relationships are then used:

$$\begin{aligned} E[(\delta_n^{(m)}(k+1))^2 \mid S_N(k)] &= E[\delta_n^{(m)}(k+1) \mid S_N(k)] \\ &= E[S_N^{(m)}(k+1) \mid S_N(k)] \\ &= V^{(m)}(S_N(k)), \end{aligned} \quad [1.3.24]$$

and also

$$D[S_N^{(m)}(k+1) \mid S_N(k)] = \frac{D[\delta_N^{(m)}(k+1) \mid S_N(k)]}{N}. \quad [1.3.25]$$

As a result, the conditional variance of MSE [1.3.20] has the following representation:

$$\sigma_m^2(p) = D[\delta_N^{(m)}(k+1) \mid S_N(k) = p] = V^{(m)}(p)[1 - V^{(m)}(p)], \quad [1.3.26]$$

which confirms the equality [1.3.21]. $\square$

So, the MSE is characterized by a SDE, namely Proposition 1.3.4.

PROPOSITION 1.3.4.– MSEs are determined by the solutions of the SDEs

$$\Delta S_N^{(m)}(k+1) = V_0^{(m)}(S_N(k)) + \Delta\mu_N^{(m)}(k+1),$$
$$0 \leq m \leq M,\ k \geq 0. \quad [1.3.27]$$

The RFI $V_0^{(m)}(p)$ is given by the relations [1.3.9]. The martingale differences $\Delta\mu_N^{(m)}(k+1)$ are determined by the quadratic characteristics [1.3.21].

1.3.4. *Convergence to the equilibrium state*

MSEs, as the solutions of the SDE [1.3.27], are represented in the form of normalized sums of random variables

$$S_N^{(m)}(k+1) := \frac{1}{N}\sum_{n=1}^{N} \delta_n^{(m)}(k+1), \quad 0 \leq m \leq M, \quad k \geq 0, \quad [1.3.28]$$

which are divided into two components: the predicted component given by the RF [1.3.19] and the stochastic, given by the martingale differences [1.3.17]. The sampling random variables in [1.3.28] are also divided into two components:

$$\delta_n^{(m)}(k+1) = E[\delta_n^{(m)}(k+1) \mid S_N(k)] + \beta_N^{(m)}(k+1),$$
$$0 \leq n \leq N. \quad [1.3.29]$$

The first component is given by the RF [1.3.19]:

$$E[\delta_n^{(m)}(k+1) \mid S_N(k)] = V^{(m)}(S_N(k)), \quad 0 \leq m \leq M, \quad k \geq 0, \quad [1.3.30]$$

which does not depend on n, where $1 \leq n \leq N$.

The stochastic component is determined by random variables, characterized by the first two moments:

$$\ldots 0, \quad 0 \leq m \leq M, \quad k \geq 0,$$
$$\ldots^2 \mid S_N(k) = p] = \sigma_m^2(p), 0 \leq m \leq M, \quad k \geq 0 \quad [1.3.31]$$

This justifies the use of Kolmogorov's theorem (Shiryaev 2011) on the convergence, with probability 1, of normalized sums of uniformly bounded sampling random variables. In the dynamic MSE model, the induction method with the stages $k \geq 0$ is used.

THEOREM 1.3.2.– *The convergence with probability 1 of the initial conditions*

$$S_N^{(m)}(0) \xrightarrow{P1} \rho_m, \quad N \to \infty, \quad 0 \leq m \leq M,$$

provides the convergence, with probability 1, of MSEs at each finite stage $k > 0$*:*

$$S_N^{(m)}(k) \xrightarrow{P1} \rho_m, \quad N \to \infty, \quad 0 \leq m \leq M, \quad k \geq 1. \qquad [1.3.32]$$

Proof of Theorem 1.3.2.– The proof is similar to that of Theorem 1.2.2. The basic idea is to prove the convergence to equilibrium of the SDEs solution

$$\Delta S_N^{(m)}(k+1) = V_0^{(m)}(S_N(k)) + \Delta\mu_N^{(m)}(k+1), \quad k \geq 0. \qquad [1.3.33]$$

The predicted component converges with probability 1 to zero using the assumption of induction:

$$P1 \lim_{N\to\infty} V_0^{(m)}(S_N(k)) = V_0^{(m)}(\rho) = 0, \quad 0 \leq m \leq M. \qquad [1.3.34]$$

The stochastic component is determined by the averaged sum of random sample values, characterized by the first two moments [1.3.31], which converges with probability 1 to zero, and also by Kolmogorov's theorem. □

1.3.5. *Normal approximation of the stochastic component*

As in sections 1.3.1 and 1.3.2, the stochastic component, given by the martingale differences [1.3.17], is approximated by normally distributed random variables with a dispersion determined by the convergence of quadratic characteristics.

THEOREM 1.3.3.– *Under the conditions of Theorem 1.3.2, the normalized martingale differences converge in distribution, with* $N \to \infty$, *to normally distributed random variables for each fixed stage* $k \geq 0$*:*

$$\sqrt{N}\mu_N^{(m)}(k+1) \xrightarrow{D} \sigma_m W(k+1), \quad 0 \leq m \leq M, \quad k \geq 0, \quad [1.3.35]$$

with dispersions $\sigma_m^2 = \rho_m(1-\rho_m)$, $0 \leq m \leq M$.

The normally distributed random variables $W(k+1)$ are determined by the first two moments:

$$EW(k+1) = 0, \quad EW^2(k+1) = 1, \quad 0 \leq m \leq M, \quad k \geq 0, \quad [1.3.36]$$

and are jointly independent for finite $k < \infty$.

Proof of Theorem 1.3.3.– Consider the quadratic characteristic of normalized stochastic components:

$$E\left[(\Delta\mu_N^{(m)}(k+1))^2 \,|\, S_N(k)\right] = \frac{\sigma_m^2(S_N^{(m)}(k))}{N},$$

$$0 \leq m \leq M, \quad k \geq 0. \quad [1.3.37]$$

Then we use Theorem 1.3.2 to prove the convergence with probability 1 of the quadratic characteristics:

$$P1 \lim_{N\to\infty} \sigma_m^2(S_N(k)) = \sigma_m^2 = \rho_m(1-\rho_m), \quad 0 \leq m \leq M. \quad [1.3.38]$$

Then, we finally apply the central limit theorem for normalized sums of uniformly bounded sample random variables:

$$\sqrt{N}\Delta\mu_N^{(m)}(k+1) = \frac{1}{\sqrt{N}}\sum_{n=1}^{N}\beta_n^{(m)}(k+1), \quad k \geq 0, \quad [1.3.39]$$

which is characterized by the first two moments:

$$E\beta_n^{(m)}(k+1) = 0, \;\; E[(\beta_n^{(m)}(k+1))^2 \,|\, S_N(k) = p] = \sigma_m^2(p),$$

$$1 \leq n \leq N, \; 0 \leq m \leq M, \quad k \geq 0. \quad [1.3.40]$$

Theorem 1.3.3 is proved. □

The approximation of the stochastic component [1.3.35] provides the basis for the introduction of a normal MSE model.

PROPOSITION 1.3.5.– The MSE is determined by the solutions of the SDE:

$$\Delta S_N^{(m)}(k+1) = V_0^{(m)}(S_N(k)) + \frac{\sigma_m}{\sqrt{N}}W^{(m)}(k+1),$$

$$0 \leq m \leq M, \quad k \geq 0, \quad [1.3.41]$$

$\sigma_m^2 = \rho_m(1-\rho_m)$, $0 \leq m \leq M$; here the normally d random variables $W_m(k+1)$, $0 \leq m \leq M$, are jointly nite $k < \infty$.

1.4. SEs with Wright–Fisher normalization

The SEs with Wright–Fisher normalization have been developed in population genetics studying the genetic code interactions (Ethier and Kurtz 1986, Ch. 10; Koroliouk and Koroliuk 2019), formalized as EPs of probability frequencies. In this section, we start from the RFs generating the EPs, given by relations of quadratic forms, and calculate the RFs of increments and introduce the linear component of the fluctuations, respectively, to equilibrium points (Proposition 1.4.1). The RFs of increments for MSEs, given in Propositions 1.3.2 and 1.3.3, provide their convergence to the equilibrium state and normal approximation of stochastic components, as $N \to \infty$.

1.4.1. *Binary RFs*

The binary EPs $P_\pm(k)$, $k \geq 0$, are given by conditional mathematical expectations

$$P_\pm(k+1) = E[S_N^\pm(k+1) \,|\, S_N^\pm(k) = P_\pm(k)], \quad k \geq 0. \qquad [1.4.1]$$

The binary RF [1.4.1] is characterized by the ratio of quadratic forms

$$P_\pm(k+1) = \frac{W_\pm(P_\pm(k))}{W(P(k))}, \quad k \geq 0, \qquad [1.4.2]$$

$$W_\pm(p_\pm) = p_\pm(W_\pm p_\pm + p_\mp), \qquad [1.4.3]$$

$$W(p) = W_+(p_+) + W_-(p_-) = W_+ p_+^2 + 2p_+p_- + W_- p_-^2. \qquad [1.4.4]$$

The probabilities of frequencies satisfy the usual conditions $0 \leq p_\pm \leq 1$ $(p_+ + p_- = 1)$ and $0 < W_\pm < 1$.

For increment probabilities,

$$\Delta P_\pm(k+1) := P_\pm(k+1) - P_\pm(k), \quad k \geq 0,$$

the corresponding RFIs are determined by

$$W_0^\pm(p_\pm) = \frac{V_0^\pm(p_\pm)}{W(p)}, \qquad [1.4.5]$$

$$V_0^\pm(p_\pm) = W_\pm(p_\pm) - p_\pm W(p). \qquad [1.4.6]$$

Taking into account the representation of the denominator [1.4.4], the numerator is given by

$$V_0^{\pm}(p_{\pm}) = p_{\mp} W_{\pm}(p_{\pm}) - p_{\pm} W_{\mp}(p_{\mp}). \qquad [1.4.7]$$

Further calculation of RFIs [1.4.5]–[1.4.7] uses the drift parameters $V_{\pm} := 1 - W_{\pm}$ and the equilibria $\rho_{\pm} := V_{\mp}/V$ $(V := V_+ + V_-)$. Then the numerator [1.4.7] is transformed using [1.4.3] in the following form:

$$W_{\pm}(p_{\pm}) = p_{\pm}(1 - V_{\pm} p_{\pm}), \qquad [1.4.8]$$

and also

$$W(p) = 1 - (V_+ p_+^2 + V_- p_-^2) = 1 - V(\rho_- p_+^2 + \rho_+ p_-^2). \qquad [1.4.9]$$

According to the formulae [1.4.7]–[1.4.9],

$$V_0^{\pm}(p_{\pm}) = V p_+ p_- (\rho_{\pm} p_{\mp} - \rho_{\mp} p_{\pm}). \qquad [1.4.10]$$

The linear component [1.4.10] can be represented as the fluctuations:

$$\rho_+ p_- - \rho_- p_+ = (p_- - \rho_-) = -(p_+ - \rho_+). \qquad [1.4.11]$$

So, the RFI numerator [1.4.10] has a representation

$$V_0^{\pm}(p_{\pm}) = -V p_+ p_- (p_{\pm} - \rho_{\pm}), \qquad [1.4.12]$$

which coincides with the RFIs of the binary EPs described in section 1.2.1.

The denominator has a representation

$$W(p) = 1 - V\pi(p_+^2/\rho_+ + p_-^2 \rho_-), \quad \pi = \rho_+ \rho_-, \qquad [1.4.13]$$

$$W(\rho) = 1 - V\pi, \quad 0 < W(\rho) < 1. \qquad [1.4.14]$$

There are obvious identities:

$$p_+ + p_- = 1, \quad \rho_+ + \rho_- = 1, \qquad [1.4.15]$$

ıl balance condition:

$$V_0^-(p_-) \equiv 0. \qquad [1.4.16]$$

1.4.2. *Multivariate RFIs*

The EP of the probability frequency of a genetic code frequency at each stage $k \geq 0$, assuming a finite number of values in a set $E = \{e_0, e_1, \dots, e_M\}$, $M \geq 1$, is given by the RF (Koroliouk *et al.* 2014):

$$P_m(k+1) := \frac{W_m(p)}{W(p)}, \quad 0 \leq m \leq M, \quad k \geq 0, \qquad [1.4.17]$$

$$W_m(p) := p_m \sum_{n=0}^{M} W_{mn} p_n, \quad 0 \leq m \leq M, \qquad [1.4.18]$$

$$W(p) := \sum_{m=0}^{M} W_m(p). \qquad [1.4.19]$$

The probabilistic frequencies have obvious limitations $0 \leq p_m \leq 1$, $\sum_{n=0}^{M} p_m = 1$. The survival parameters also have limitations $0 \leq W_{mn} \leq 1$, $0 \leq m, n \leq M$.

For increment probabilities,

$$\Delta P_m(k+1) := P_m(k+1) - P_m(k), \quad 0 \leq m \leq M, \quad k \geq 0, \qquad [1.4.20]$$

the corresponding RFIs have the form

$$W_0^{(m)}(p) = \frac{V_0^{(m)}(p)}{W(p)}, \quad 0 \leq m \leq M, \qquad [1.4.21]$$

$$V_0^{(m)}(p) = W_m(p) - p_m W(p), \quad 0 \leq m \leq M. \qquad [1.4.22]$$

Using equivalently modified drift parameters

$$V_{mn} := 1 - W_{mn}, \quad 0 \leq m, n \leq M, \qquad [1.4.23]$$

and the formula [1.4.22], we have a representation of the numerator of [1.4.21]:

$$V_0^{(m)}(p) = p_m \left[\sum_{n=0}^{M} p_n (V_n \cdot p) - (V_m \cdot p) \right], \quad 0 \leq m \leq M, \qquad [1.4.24]$$

and the normalizing denominator of [1.4.21]:

$$W(p) = 1 - \sum_{n=0}^{M} p_m(V_m \cdot p), \quad 0 \le m \le M. \qquad [1.4.25]$$

The scalar products

$$(V_m \cdot p) := \sum_{n=0}^{M} V_{mn} p_n, \quad 0 \le m \le M,$$

characterize the fluctuations.

In particular, for the diagonal matrix of drift parameters

$$\mathbb{V} = [V_m \delta_{mn}, \quad 0 \le m, n \le M], \qquad [1.4.26]$$

the equilibrium state is defined as

$$\rho_m = V_m^{-1}, \quad 0 \le m \le M. \qquad [1.4.27]$$

Under the additional condition on the drift parameter normalization

$$\sum_{m,n=0}^{M} \overline{V}_{mn} = 1,$$

the equilibria have the following representation:

$$\rho_m = \overline{V}_m = \sum_{n=0}^{M} \overline{V}_{mn}, \quad 0 \le m \le M, \qquad [1.4.28]$$

where the elements of the inverse matrix are summarized

$$\mathbb{V}^{-1} = [\overline{V}_{mn}, \quad 0 \le m, n \le M],$$

relative to the drift matrix

$$\mathbb{V} = [V_{mn}, \quad 0 \le m, n \le M], \quad 0 \le m \le M,$$

under the additional condition of normalization

$$\sum_{m=0}^{M} \overline{V}_m = \sum_{m,n=0}^{M} \overline{V}_{mn} = 1.$$

The representation of scalar products of the fluctuations using the relation

$$(V_m \cdot p) = \frac{V_0 \pi p_m}{\rho_m}, \quad 0 \le m \le M, \qquad [1.4.29]$$

is postulated in the following proposition.

PROPOSITION 1.4.1.– (*Basic assumption 2*). The numerators of RFI [1.4.24] and the normalizing denominator [1.4.25] are given by the relation of the fluctuations [1.4.30] and [1.4.31]:

$$V_0^{(m)}(p) = V_0 \pi p_m \left[\sum_{n=0}^{M} p_n^2/\rho_n - p_m/\rho_m \right], \quad 0 \le m \le M. \qquad [1.4.30]$$

The normalizing denominator has the form

$$W(p) = 1 - V_0 \pi \sum_{n=0}^{M} p_n^2/\rho_n. \qquad [1.4.31]$$

REMARK 1.4.1.– The RFI [1.4.30] can be transformed using the identity $\sum_{l=0}^{M} p_l = 1$:

$$V_0^{(m)}(p) = V_0 \pi p_m \sum_{n=0}^{M} p_n \left[p_n/\rho_n - p_m/\rho_m \right], \quad 0 \le m \le M, \qquad [1.4.32]$$

with linear components

$$\left[p_n/\rho_n - p_m/\rho_m \right], \quad 0 \le m, n \le M,$$

which satisfy the local balance condition for the RFIs:

$$V_0^{(m)}(\rho) \equiv 0, \quad 0 \le m \le M.$$

1.5. Exponential statistical experiments

Consider *exponential statistical experiments* (ESEs), given by sample random variables $(\delta_n(k),\ 1 \le n \le N)$, $k \ge 0$. For ESEs in a series scheme $(N \to \infty)$, a steady regime is established, which is determined by the RF equilibrium, and can be approximated by a process of exponential Brownian motion.

ESEs are widely known in financial mathematics (e.g., *(B, S)* markets; Shiryaev 1999, Ch. 6, section 4).

1.5.1. *Binary ESEs*

Recall the definition of binary SEs (see sections 1.1 and 1.2), as averaged sums of random sample variables $\delta_n(k)$, $1 \leq n \leq N$, $k \geq 0$, which take two possible values ± 1:

$$S_N(k) := \frac{1}{N} \sum_{n=1}^{N} \delta_n(k), \quad k \geq 0. \qquad [1.5.1]$$

The dynamics of SEs, using $k \geq 0$, are determined by the RF:

$$E[\delta_n(k+1) \mid S_N(k) = c] = V(c), \quad |c| \leq 1, \quad k \geq 0. \qquad [1.5.2]$$

A linear RF has the form

$$\begin{aligned} V(c) &= (1 - V)c + V\rho = c - V(c - \rho) \\ &= \rho + (1 - V)(c - \rho), \ |c| \leq 1. \end{aligned} \qquad [1.5.3]$$

Here, as usual, ρ denotes the equilibrium state of the RF:

$$V(\rho) = \rho. \qquad [1.5.4]$$

The RFs [1.5.2] and [1.5.3] define the predicted component of the binary SEs:

$$C(k+1) := E[S_N(k+1) \mid S_N(k) = C(k)] = V(C(k)), \quad k \geq 0. \quad [1.5.5]$$

Also, binary SEs [1.5.1]–[1.5.3] are characterized by the conditional dispersion:

$$D[S_N(k+1) \mid S_N(k) = c] = \frac{\sigma^2(c)}{N}, \ \sigma^2(c) := 1 - V^2(c), \ |c| \leq 1. \ [1.5.6]$$

DEFINITION 1.5.1.– ESEs are given by random variables $\delta_n(k)$, $1 \leq n \leq N$, $k \geq 0$, which take two values ± 1, as follows:

$$\mathcal{E}_N(k, \lambda) := \prod_{n=1}^{N} [1 + \lambda \delta_n(k)], \quad k \geq 0. \qquad [1.5.7]$$

The sample random variables $\delta_n(k)$, $1 \leq n \leq N$, $k \geq 0$, are jointly independent at fixed $k \geq 0$ and equally distributed at different n, where $1 \leq n \leq N$.

The conditional mathematical expectation of ESEs has a representation

$$\Pi_N(k,\lambda) := E[\mathcal{E}_N(k+1,\lambda) \mid S_N(k)] = [1+\lambda V(S_N(k))]^N, \quad [1.5.8]$$

with the RF $V(c)$ [1.5.3]. The drift parameter V and the equilibrium state ρ satisfy the following conditions:

$$0 < V < 2, \quad |\rho| \leq 1. \quad [1.5.9]$$

1.5.2. *Steady regime of ESEs*

ESE [1.5.7] with RF [1.5.3] in the series scheme with the parameter $N \to \infty$ are investigated with the normalization $\lambda_N := \lambda/N$. So,

$$\mathcal{E}_N(k,\lambda/N) := \prod_{n=1}^{N}[1+\lambda\delta_n(k)/N], \quad k \geq 0, \quad [1.5.10]$$

and the normalized average is

$$\Pi_N(k,\lambda/N) = [1+\lambda V(S_N(k))/N]^N. \quad [1.5.11]$$

Introducing an *exponential martingale*

$$\mu_N^e(k+1,\lambda/N) := \mathcal{E}_N(k+1,\lambda/N)\,/\,\Pi_N(k,\lambda/N), \quad k \geq 0, \quad [1.5.12]$$

whose conditional expectation is

$$E[\mu_N^e(k+1,\lambda/N) \mid S_N(k)] = 1, \quad k \geq 0. \quad [1.5.13]$$

The steady regime of ESEs sets the following theorem.

THEOREM 1.5.1.– *Under the condition of initial values, convergence with probability 1:*

$$S_N(0) \xrightarrow{P1} \rho, \quad N \to \infty, \quad [1.5.14]$$

a convergence in probability takes place, of ESEs in the series scheme as $N \to \infty$:

$$\mathcal{E}_N(k,\lambda/N) \xrightarrow{P} \exp(\lambda\rho), \quad N \to \infty, \quad k \geq 0, \quad [1.5.15]$$

and also the convergence of normalized averages:

$$\Pi_N(k,\lambda/N) \xrightarrow{P} \exp(\lambda\rho), \quad N \to \infty, \quad k \geq 0. \quad [1.5.16]$$

COROLLARY 1.5.1.– *In conditions* [1.5.14], *there is a convergence of the exponential martingale:*

$$\mu_N(k+1, \lambda/N) \xrightarrow{P} 1, \quad N \to \infty, \quad k \geq 0. \qquad [1.5.17]$$

Proof of Theorem 1.5.1.– The Le Cam formula of approximation is used (Borovskikh and Korolyuk 1997, Lemma 6.3.1).

LEMMA 1.5.1.– *Under the condition of sampling random variables convergence,*

$$\max_{1 \leq n \leq N} |\delta_n(k+1)/N| \xrightarrow{P} 0, \; N \to \infty \quad \forall k \geq 0, \qquad [1.5.18]$$

a convergence in probability takes place

$$\sum_{n=1}^{N} \ln[1 + \lambda\delta_n(k+1)/N] - \lambda S_N(k+1) \xrightarrow{P} 0, \; N \to \infty. \qquad [1.5.19]$$

For binary random sampling variables $\delta_n(k)$, $1 \leq n \leq N$, $k \geq 0$, the condition of convergence [1.5.18] is fulfilled. According to Theorem 1.1.1 (section 1.1), a convergence with probability 1 takes place:

$$S_N(k+1) \xrightarrow{P1} \rho, \; N \to \infty. \qquad [1.5.20]$$

Note that the convergence in probability [1.5.19] implies the following:

$$P \cdot \lim_{N\to\infty} \mathcal{E}_N(k+1, \lambda/N)$$

$$= P \cdot \lim_{N\to\infty} \exp \sum_{n=1}^{N} \ln[1 + \lambda\delta_n(k+1)/N] = \exp(\lambda\rho). \qquad [1.5.21]$$

Similarly (with obvious simplifications), the convergence in probability [1.5.16] for normalized averages [1.5.11] can be established. Theorem 1.5.1 is proved. □

1.5.3. *Approximation of ESEs by geometric Brownian motion*

ESEs in the scheme of approximation by an exponential stochastic process are scaled with normalizing $\lambda_N = \lambda/\sqrt{N}$, $N \to \infty$:

$$\mathcal{E}_N(k, \lambda/\sqrt{N}) = \prod_{n=1}^{N} [1 + \lambda\delta_n(k)/\sqrt{N}], \quad k \geq 0. \qquad [1.5.22]$$

Consequently, the normalized average is

$$\Pi_N(k, \lambda/\sqrt{N}) = [1 + \lambda V(S_N(k))/\sqrt{N}]^N, \quad k \geq 0. \qquad [1.5.23]$$

The exponential martingale

$$\mu_N^e(k+1, \lambda/\sqrt{N}) := \mathcal{E}_N(k+1, \lambda/\sqrt{N})/\Pi_N(k, \lambda/\sqrt{N}), \quad k \geq 0, \qquad [1.5.24]$$

retains the martingale property [1.5.13].

THEOREM 1.5.2 (*Approximation of ESEs*).– *In the conditions of Theorem 1.5.1, there is a convergence in distribution* ($\sigma^2 := 1 - \rho^2$)*:*

$$\mu_N^e(k, \lambda/\sqrt{N}) \xrightarrow{d} \exp[\lambda\sigma W(k) - \lambda^2\sigma^2/2], \; N \to \infty, \quad k \geq 1. \quad [1.5.25]$$

REMARK 1.5.1.– The limit exponential stochastic process

$$\exp[\lambda\sigma W(k) - \lambda^2\sigma^2/2], \quad k \geq 0, \qquad [1.5.26]$$

is a process of exponential Brownian motion (Shiryaev 1999), which plays an important role in modern financial mathematics.

Proof of Theorem 1.5.2.– As before, we used Le Cam approximation, but now in other initial conditions.

LEMMA 1.5.2.– *(Le Cam Approximation (Borovskikh and Korolyuk 1997, Lemma 6.3.1)). Let the convergence in probability take place, of normalized random variables:*

$$\max_{1 \leq n \leq N} |\delta_n(k+1)/\sqrt{N}| \xrightarrow{P} 0, \; N \to \infty, \qquad [1.5.27]$$

and let the sums

$$V_N(k) := \frac{1}{N} \sum_{n=1}^{N} \delta_n^2(k), \quad k \geq 0, \qquad [1.5.28]$$

be bounded in probability.

Then the convergence in probability takes place:

$$\sum_{n=1}^{N} \ln[1 + \lambda\delta_n(k+1)/\sqrt{N}] - \lambda\sqrt{N}S_N(k+1)$$
$$+ \lambda^2 V_N(k+1)/2 \xrightarrow{P} 0, \quad N \to \infty. \qquad [1.5.29]$$

REMARK 1.5.2.– Consider the case of the binary sample variables $\delta_n(k)$, $1 \le n \le N$, $k \ge 0$, which take two values ± 1. So the sum [1.5.28] is equal to 1:

$$V_N(k) \equiv 1 \quad \forall k \ge 0. \qquad [1.5.30]$$

The convergence with probability [1.5.29] is equivalent to the following convergence, in probability, as $N \to \infty$:

$$\mathcal{E}_N(k+1,\, \lambda/\sqrt{N}) \cdot \exp\Big[-\lambda\sqrt{N}S_N(k+1)\Big] \xrightarrow{P} \exp[-\lambda^2/2],$$
$$k \ge 0. \qquad [1.5.31]$$

The conditions of Lemma 1.5.2 allow us to justify the convergence, in probability, of the normalized averages:

$$\Pi_N(k,\, \lambda/\sqrt{N}) \cdot \exp\Big[-\lambda\sqrt{N}V(S_N(k))\Big] \xrightarrow{P} \exp[-\lambda^2\rho^2/2],$$
$$N \to \infty, \quad k \ge 0. \quad [1.5.32]$$

In this case, the convergence, with probability 1, of the RFs is used (Theorem 1.1.1):

$$V^2(S_N(k)) \xrightarrow{P1} V^2(\rho) = \rho^2, \quad N \to \infty, \quad k \ge 0. \qquad [1.5.33]$$

In order to substantiate the convergence [1.5.32], the Taylor decomposition of the logarithmic function is used, up to the third term (see [1.5.23]):

$$\begin{aligned}\Pi_N(k,\, \lambda/\sqrt{N}) &= \exp N \ln[1 + \lambda V(S_N(k))/\sqrt{N}] \\ &= \exp \lambda\sqrt{N}V(S_N(k)) + \frac{\lambda^2}{2}V^2(S_N(k)) \\ &\quad + \frac{\lambda^3}{6\sqrt{N}}R_N(S_N(k)), k \ge 0, \end{aligned} \qquad [1.5.34]$$

with a neglected summand $R_N(S_N(k)) \to 0$, $N \to \infty$.

The convergence in distribution of the exponential martingale [1.5.25] is justified by using the convergences [1.5.31] and [1.5.32] in the representation of the exponential martingale:

$$\mu_N^e(k+1,\, \lambda/\sqrt{N}) = \exp\{\lambda\sqrt{N}[S_N(k+1) - V(S_N(k))]\}$$
$$\times \frac{\mathcal{E}_N(k+1,\, \lambda/\sqrt{N})\exp[-\lambda\sqrt{N}S_N(k+1)]}{\Pi_N(k+1,\, \lambda/\sqrt{N})\exp[-\lambda\sqrt{N}V(S_N(k))]}. \quad [1.5.35]$$

The first factor [1.5.35] can be transformed as

$$\sqrt{N}[S_N(k+1) - V(S_N(k))] = \sqrt{N}[\Delta S_N(k+1) + V_0(S_N(k))]$$
$$= \sqrt{N}\Delta\mu_N(k+1), \quad k \geq 0. \quad [1.5.36]$$

According to Theorem 1.2.3, the martingale differences are approximated by a Gaussian process:

$$\sqrt{N}[S_N(k+1) - V(S_N(k))]$$
$$= \sqrt{N}\Delta\mu_N(k+1) \xrightarrow{d} \sigma W(k+1), \quad k \geq 0, \quad \sigma^2 = 1 - \rho^2. \quad [1.5.37]$$

The convergences [1.5.31] and [1.5.32], together with [1.5.37], prove the convergence, by distribution, of the exponential martingale [1.5.25] to an exponential Brownian motion [1.5.26]. Theorem 1.5.2 is proved. □

The limit Theorem 1.5.2 and the formula [1.5.24] for the exponential martingale give an opportunity to represent ESE [1.5.10] in the series scheme in the following form:

$$\mathcal{E}_N(k+1, \lambda/N) = \Pi_N(k, \lambda/N)\mu_N^e(k+1, \lambda/N). \quad [1.5.38]$$

Now we use the asymptotic relations for the multipliers in [1.5.38]. In the sequel, the residual term R_N is understood as $R_N \to 0$ by $N \to \infty$.

$$\Pi_N(k, \lambda/N) \overset{\mathrm{d}}{=} \exp[\lambda V(S_N(k)) - \lambda^2\rho^2/2N]e^{R_N}.$$

$$\mu_N^e(k+1, \lambda/N) \overset{\mathrm{d}}{=} \exp[\lambda(\sigma/\sqrt{N})W(k+1) - \lambda^2\sigma^2/2N]e^{R_N}.$$

As the result, we obtain the following asymptotic representation of ESEs:

$$\mathcal{E}_N(k+1,\, \lambda/N) \overset{\mathrm{d}}{\approx} \exp[\lambda V(S_N(k)) - \lambda^2\rho^2/2N]$$
$$\times \exp[\lambda(\sigma/\sqrt{N})W(k+1) - \lambda^2\sigma^2/2N]. \quad [1.5.39]$$

Otherwise, taking into account [1.5.37], we obtain the following approximation:

$$\mathcal{E}_N(k+1, \lambda/N) \overset{\mathrm{d}}{\approx} \exp\{\lambda[V(S_N(k)) \\ + (\sigma/\sqrt{N})W(k+1)] - \lambda^2/2N\}. \quad [1.5.40]$$

The asymptotic formulae [1.5.39] and [1.5.40] give a basis for further approximation of ESEs by the *normal process of autoregression.*

PROPOSITION 1.5.1.– The ESEs [1.5.10] are approximated by the normal process of autoregression

$$\prod_{n=1}^{N}[1+\lambda\widetilde{\delta}_n(k+1)/N] = \exp[\lambda V(\widetilde{S_N}(k)) - \lambda^2\rho^2/2N] \\ \times \exp[\lambda(\sigma/\sqrt{N})W(k+1) - \lambda^2\sigma^2/2N], \quad [1.5.41]$$

or, that the same thing,

$$\prod_{n=1}^{N}[1+\lambda\widetilde{\delta}_n(k+1)/N] \\ = \exp\{\lambda[V(\widetilde{S_N}(k)) + (\sigma/\sqrt{N})W(k+1)] - \lambda^2/2N\}. \quad [1.5.42]$$

REMARK 1.5.3.– An important basis for the approximation, by the normal process of autoregression [1.5.41] or [1.5.42], is that the conditional mathematical expectation of the normal process of autoregression [1.5.41] (or [1.5.42]) asymptotically converges to the RF (conditional mathematical expectation) of the initial ESE [1.5.11], namely

$$E\left[\prod_{n=1}^{N}[1+\lambda\delta_n(k+1)/N]|S_N(k)\right] = \exp[\lambda V(S_N(k)) - \lambda^2\rho^2/2N] \\ = \Pi_N(k, \lambda/N)e^{-R_N}. \quad [1.5.43]$$

Consequently, the approximation by a normal process of autoregression [1.5.41] (or [1.5.42]) of ESEs using the exponential Brownian motion process [1.5.26], obtained in Theorem 1.5.2, creates a variety of perspectives in applications of models of the SEs [1.5.11] in economics and biology.

2

Diffusion Approximation of Statistical Experiments in Discrete–Continuous Time

The discrete Markov processes (DMPs), given by the solutions of the stochastic difference equations (SDEs) in discrete–continuous time, are approximated (as $N \to \infty$) by a diffusion process with evolution. The DMPs are also approximated in a Markov random environment (MRE; sections 2.3 and 2.4). The DMPs, considered as special semimartingales with random time change, are explored in section 2.5. In section 2.7, the DMPs with asymptotically small diffusion (ASD) are characterized by exponential generators (EGs) and action functionals (AFs), which correspond to the diffusion Ornstein–Uhlenbeck processes.

The discrete–continuous time is determined by the connection of discrete instants of time $k \geq 0$, $k \in \mathbb{N} := \{0, 1, ...\}$, with continuous time $t \geq 0$, $t \in \mathbb{R}_+ := [0, +\infty)$, by the formula

$$k = [Nt], \quad t \geq 0. \qquad [2.0.1]$$

The integer part of the product means that

$$[Nt] = k \quad \text{for} \quad t_k^N := k/N, \quad k \geq 0. \qquad [2.0.2]$$

The transition from the discrete instants of time $k \geq 0$ to the continuous time $t \geq 0$ means that with increasing the sample size $N \to \infty$, the intervals between adjacent moments of continuous time tend to 0:

$$\Delta_N := t_{k+1}^N - t_k^N = 1/N \to 0, \quad N \to \infty. \qquad [2.0.3]$$

However, the condition

$$t_k^N = k/N \to t, \quad N \to \infty, \qquad [2.0.4]$$

also means that $k \to \infty$.

2.1. Binary DMPs

Binary *statistical experiments* (SEs) are defined by averaged sums of sample variables $\delta_n(k)$, $1 \le n \le N$, $k \ge 0$, jointly independent for fixed $k \ge 0$ and identically distributed for different $n \in [1, N]$, taking two values ± 1 (see section 1.1):

$$S_N(k) := \frac{1}{N}\sum_{n=1}^{N} \delta_n(k), \quad -1 \le S_N(k) \le 1, \quad k \ge 0. \qquad [2.1.1]$$

The binary SE [2.1.1] can also specify the difference between the frequency SEs (see section 1.1):

$$S_N(k) = S_N^+(k) - S_N^-(k), \quad k \ge 0. \qquad [2.1.2]$$

In the general case, both values are arbitrary $\{e_0, e_1\}$, the binary SE [2.1.1] can be considered in the vector form:

$$S_N(k) = (S_N^+(k), S_N^-(k)),\ k \ge 0. \qquad [2.1.3]$$

Introducing *evolutionary processes* (EPs), given by the conditional mathematical expectations:

$$P_\pm(k+1) := E[S_N^\pm(k+1) \mid S_N^\pm(k) = P_\pm(k)], \quad k \ge 0. \qquad [2.1.4]$$

Obviously, we obtain the global balance condition

$$P_+(k) + P_-(k) \equiv 1 \quad \forall k \ge 0. \qquad [2.1.5]$$

The dynamics of the EP [2.1.4] is determined for increments

$$\Delta P_\pm(k+1) = P_\pm(k+1) - P_\pm(k), \quad k \ge 0, \qquad [2.1.6]$$

which evidently also satisfies the global balance condition

$$\Delta P_+(k) + \Delta P_-(k) \equiv 0 \quad \forall k \ge 0. \qquad [2.1.7]$$

The EP increments [2.1.6] are given by the conditional mathematical expectations:

$$V_0^{\pm}(p_{\pm}) = E[\Delta S_N^{\pm}(k+1) \mid S_N^{\pm}(k) = p_{\pm}], \quad k \geq 0. \qquad [2.1.8]$$

The *regression function of increments* (RFI) [2.1.8] also satisfy the global balance condition

$$V_0^+(p_+) + V_0^-(p_-) \equiv 0 \quad \forall p_{\pm}, \; p_+ + p_- = 1. \qquad [2.1.9]$$

The choice of RFI [2.1.8] and the corresponding EP [2.1.4] is a fundamental problem in studying the dynamics of the EP and is realized in a form of basic assumptions.

The basic principle of the RFI [2.1.8] selection is based on the *linear dependence* of the RFIs from the fluctuations:

$$\widehat{P}_{\pm}(k) := P_{\pm}(k) - \rho_{\pm},$$

where $\rho_{\pm}$ are the equilibrium states: $V_0^{\pm}(\rho_{\pm}) = 0$. However, the nonlinear component of the RFI should provide a *balanced impact* on the dynamics of the EP [2.1.4].

Basic assumption 2.1.1. The RFI of a binary EP [2.1.4] is given by

$$V_0^{\pm}(p_{\pm}) := -V_0 p_+ p_-(p_{\pm} - \rho_{\pm}). \qquad [2.1.10]$$

The frequency EP [2.1.4] is given by the *difference evolutionary equation* (DEE) solutions

$$\Delta P_{\pm}(k+1) = -V_0^{\pm}(P_{\pm}(k)), \quad k \geq 0. \qquad [2.1.11]$$

2.1.1. *DMPs in discrete–continuous time*

The study of SEs in section 2.1 justifies the introduction of DMPs in discrete–continuous time:

$$\zeta_N^{\pm}(t) := \sqrt{N}[S_N^{\pm}(k) - \rho_{\pm}], \quad k = [Nt], \quad t \geq 0. \qquad [2.1.12]$$

The normalized fluctuations of SE [2.1.12] mean that

$$S_N^{\pm}(k) = \rho_{\pm} + \zeta_N^{\pm}(t)/\sqrt{N}, \quad t \geq 0. \qquad [2.1.13]$$

The global balance condition for the DMP [2.1.12] is also evident

$$\zeta_N^+(t) + \zeta_N^-(t) \equiv 0 \quad \forall t \geq 0. \qquad [2.1.14]$$

Thus, the DMP [2.1.12] is uniquely determined by one of the components $\zeta_N^\pm(t)$, $t \geq 0$.

In the sequel, we will use the notation $\zeta(t) = \zeta^+(t)$.

The dynamics of the DMP [2.1.12] is determined by the solution of the SDEs given by two components: the predictable component defined by RFI [2.1.10] and the stochastic component with the normalized quadratic characteristic.

Basic assumption 2.1.2. The DMP [2.1.12] is determined by the solution of the SDEs

$$\Delta\zeta_N(t) = -\frac{1}{N}V_0 S_N^+(k) S_N^-(k)\zeta_N(t) + \frac{1}{\sqrt{N}}\sigma(S_N^\pm(k))\Delta\mu_N(t),$$

$$t \geq 0. \quad [2.1.15]$$

The quadratic characteristic (see Lemma 1.2.1) has the form

$$\sigma^2(p) = V^+(p_+)V^-(p_-). \qquad [2.1.16]$$

The regression functions (RFs) are expressed as

$$\begin{aligned} V^\pm(p_\pm) &= p_\pm - V_0 p_+ p_- (p_\pm - \rho_\pm) \\ &= \rho_\pm + (1 - V_0 p_+ p_-)(p_\pm - \rho_\pm). \end{aligned} \qquad [2.1.17]$$

The SDE [2.1.15] is basic in the asymptotic analysis of the DMP [2.1.12] in discrete–continuous time.

The nonlinear factors of predictable component and quadratic characteristic in discrete–continuous time converge, in probability, as follows (see Lemma 2.1.3):

$$V_0(S_N(k)) := V_0 S_N^+(k) S_N^-(k) \xrightarrow{P} V_0\rho_+\rho_-, \quad k \to \infty, \qquad [2.1.18]$$

$$\sigma^2(S_N(k)) \xrightarrow{P} \rho_+\rho_-, \quad k \to \infty. \qquad [2.1.19]$$

The convergences [2.1.18] and [2.1.19] of the SDE components with nonlinear RFIs generate a linear approximation of the SDE [2.1.15].

Basic assumption 2.1.3. The DMP [2.1.12] is approximated by the solution of the next linear SDE

$$\Delta\zeta_N^0(t) = -\frac{1}{N}V_0\pi\zeta_N^0(t) + \frac{1}{\sqrt{N}}\breve{\sigma}\Delta\mu_N^0(t), \quad t \geq 0, \qquad [2.1.20]$$

where $\pi := \rho_+\rho_- = \breve{\sigma}^2$.

The DMP $\Delta\zeta_N^0(t), t \geq 0$, given by the solution of the SDE [2.1.20], can be approximated, as $N \to \infty$, in discrete–continuous time, by a *diffusion process with evolution.*

THEOREM 2.1.1.– *The DMP* $\Delta\zeta_N^0(t)$, $t \geq 0$, *determined by the solution of the SDE* [2.1.20], *converges in distribution:*

$$\zeta_N^0(t) \xrightarrow{D} \zeta^0(t), \quad N \to \infty, \quad 0 \leq t \leq T, \qquad [2.1.21]$$

to a diffusion process with evolution $\zeta^0(t)$, $t \geq 0$, *given by the following generator which acts on the class of finite real-valued functions* $\varphi(c) \in C^3(R)$ $(R := (-\infty, +\infty))$, *continuously differentiated three times with bounded derivatives:*

$$L^0\varphi(c) = -V_0\pi c\varphi'(c) + 1/2\breve{\sigma}^2\varphi''(c), \qquad [2.1.22]$$

by the additional initial conditions

$$\zeta_N^0(0) \xrightarrow{P} \zeta^0(0), \quad N \to \infty, \quad E|\zeta_N^0(t)| < +\infty. \qquad [2.1.23]$$

The diffusion process with evolution $\zeta^0(t)$, $t \geq 0$, *is an Ornstein–Uhlenbeck process with continuous time, determined by the solution of the SDEs*

$$\mathrm{d}\zeta^0(t) = -V_0\pi\zeta^0(t)\mathrm{d}t + \breve{\sigma}\mathrm{d}W(t), \quad t \in [0, T]. \qquad [2.1.24]$$

The Brownian motion $W(t)$, $t \geq 0$, *has the standard parameters*

$$E\mathrm{d}W(t) = 0, \quad E(\mathrm{d}W(t))^2 = \mathrm{d}t. \qquad [2.1.25]$$

2.1.2. *Justification of diffusion approximation*

The limit theorems for Markov random processes are based on the operator characterization of the Markov process (MP; by means of a generator) on a class of real-valued functions with argument in the set of the MP values (Korolyuk and Limnios 2005). The generators convergence on a sufficiently rich class of real-valued functions ensures the convergence of finite-dimensional distributions (Ethier and Kurtz 1986; Skorokhod 1987).

Following the monograph (Skorokhod 1987; see also Korolyuk and Limnios 2005), a generator of the DMPs with the linear RFI [2.1.20] is defined as follows:

$$L_N^0 \varphi(c) = NE[\varphi(c + \Delta\zeta_N^0(t)) - \varphi(c) \mid \zeta_N^0(t) = c]. \qquad [2.1.26]$$

Proof of Theorem 2.1.1.– An essential stage of the proof of Theorem 2.1.1 is contained in the following lemma.

LEMMA 2.1.1.– *There is a convergence of generators* [2.1.26]*:*

$$\lim_{N\to\infty} L_N^0 \varphi(c) = L^0 \varphi(c), \quad \varphi(c) \in C^3(R), \qquad [2.1.27]$$

in the class $C^3(R)$ of real-valued finite functions, three times continuously differentiable with bounded derivatives. The limit generator

$$L^0 \varphi(c) = -V_0 \pi c \varphi'(c) + 1/2 \breve{\sigma}^2 \varphi''(c), \quad \pi = \breve{\sigma}^2 = \rho_+ \rho_-, \qquad [2.1.28]$$

defines the limit process of Ornstein–Uhlenbeck-type [2.1.24].

Proof of Lemma 2.1.1.– Using representation [2.1.20] of the DMP of the fluctuations $\zeta_N^0(t)$, $t \geq 0$, let us calculate the first two moments of increments:

$$E[\Delta\zeta_N^0(t) | \zeta_N^0(t) = c] = -V_0 \pi c / N, \qquad [2.1.29]$$

$$E[(\Delta\zeta_N^0(t))^2 | \zeta_N^0(t) = c] = \breve{\sigma}^2 / N + o(1/N) \quad \forall c, \quad \breve{\sigma}^2 = \rho_+ \rho_-. \qquad [2.1.30]$$

By applying the Taylor formula in representation [2.1.26] of a generator L_N^0, using a test function $\varphi(c) \in C^3(R)$, we obtain

$$L_N^0 \varphi(c) = N\Big[E[\Delta\zeta_N^0(t) \mid \zeta_N^0(t) = c]\varphi'(c) + E[(\Delta\zeta_N^0(t))^2 \mid \zeta_N^0(t) = c]\frac{1}{2}\varphi''(c)\Big] + R_N \varphi(c). \qquad [2.1.31]$$

Here the residual term, under the condition of Lemma 2.1.1, is negligible:

$$R_N \varphi(c) \to 0, \quad N \to \infty, \quad \varphi(c) \in C^3(R). \qquad [2.1.32]$$

Using representations [2.1.29]–[2.1.31] of the first two moments of increments, we obtain the statement [2.1.27] of Lemma 2.1.1 in an asymptotic form:

$$L_N^0\varphi(c) = L^0\varphi(c) + R_N\varphi(c).$$

The representations [2.1.29]–[2.1.31] imply the assertions of Lemma 2.1.1. □

The proof of Theorem 2.1.1 is based on Theorem 1 of Skorokhod (1987, Ch. 2, section 1), from which the convergence of finite-dimensional distributions of normalized SE fluctuations follows. □

REMARK 2.1.1.– DMP $\zeta_N^+(t), t \geq 0$, with the linear RFI:

$$V_0^+(\zeta_N(t)) = -V_0\zeta_N^+(t), \quad t \geq 0, \tag{2.1.33}$$

obviously also admits the approximation by a diffusion process with evolution, with a generator

$$L^+\varphi(c) = -V_0 c\varphi'(c) + \frac{1}{2}\breve{\sigma}^2\varphi''(c), \;\; \breve{\sigma}^2 = \rho_+\rho_-. \tag{2.1.34}$$

Similarly, the following convergence takes place.

LEMMA 2.1.2.– *There is a convergence of generators*

$$\lim_{N\to\infty} L_N^+\varphi(c) = L^+\varphi(c), \quad \varphi(c) \in C^3(R),$$

on a class of real-valued finite functions $C^3(R)$, three times continuously differentiated with bounded derivatives. The limit generator L^+ is given by the representation [2.1.34].

The diffusion approximation justification for a DMP $\zeta_N^0(t)$ with linear SDE [2.1.20] obviously provides the statement of Theorem 2.1.1 for the DMP frequency $\zeta_N^+(t)$ with the linear RFI [2.1.33].

Let us introduce the generator of the DMP given by the solution of the SDE [2.1.15] with the nonlinear components [2.1.16] and [2.1.17]:

$$L_N\varphi(c) = E[\varphi(c + \Delta\zeta_N(t)) - \varphi(c) \,|\, \zeta_N(t) = c]. \tag{2.1.35}$$

COROLLARY 2.1.1.– *The convergence in probability* [2.1.18] *and [2.1.19] of the coefficients of equation* [2.1.15] *provides the asymptotic representation of the DMP generators:*

$$L_N\varphi(c) = L^0\varphi(c) + R_N\varphi(c),$$
$$R_N\varphi(c) \to 0, \quad N \to \infty, \quad \varphi(c) \in C^3(R).$$

The convergence in probability [2.1.18] and [2.1.19] of the coefficients of equation [2.1.15] is provided by the following lemma.

LEMMA 2.1.3.– *DMP* $\zeta_N^+(t)$, $t \geq 0$, *given by the solution SDE* [2.1.15] *with uniformly bounded initial values:*

$$E|\zeta_N^+(0)| \leq C < +\infty, \quad N \to \infty, \tag{2.1.36}$$

have a uniformly bounded second moment:

$$E|\zeta_N^+(t)|^2 \leq C < +\infty, \quad 0 \leq t \leq T, \quad N \to \infty. \tag{2.1.37}$$

Proof of Lemma 2.1.2.– The following semimartingale representation is used:

$$\zeta_N^+(t) = \zeta_N^+(0) + \frac{1}{N}V_N(t) + \frac{1}{\sqrt{N}}M_N(t), \quad t \geq 0. \tag{2.1.38}$$

The evolutionary component (see [2.1.10]) has the following form:

$$V_N^+(t) = -\sum_{k=0}^{[Nt]} V_0 S_N^+(k) S_N^-(k) \zeta_N(t_k^N), \quad t_k^N := k/N. \tag{2.1.39}$$

The stochastic component is

$$M_N^+(t) = \sum_{k=0}^{[Nt]} \sigma(S_N(k)) \Delta\mu_N(k+1), \quad t \geq 0. \tag{2.1.40}$$

The quadratic characteristic (see [2.1.17]) is

$$\sigma^2(p) := V^+(p)V^-(p), \quad V^{\pm}(p) := p_{\pm} - V_0 p_+ p_-(p_{\pm} - \rho_{\pm}), \tag{2.1.41}$$

and admits the following representation:

$$\sigma^2(S_N(k)) = V^+(S_N(k))V^-(S_N(k)), \tag{2.1.42}$$

$$V^{\pm}(S_N(k)) = \rho_{\pm} + [1 - V_0 S_N^+(k) S_N^-(k)]\zeta_N(t_k^N)/\sqrt{N}. \tag{2.1.43}$$

Let us introduce, following Liptser (1994), the notation

$$\zeta_N^*(t) := \sup_{0\le s\le t} |\zeta_N(s)|, \quad t \ge 0. \tag{2.1.44}$$

Similar to Liptser (1994), we have the following estimations:

$$\sup_{0\le s\le t} |V_N(s)|^2 \le NV_0^2[\zeta_N^*(t)]^2, \quad t \ge 0, \tag{2.1.45}$$

$$\sup_{0\le s\le t} |M_N(s)|^2 \le NC_1\zeta_N^*(t), \quad t \ge 0. \tag{2.1.46}$$

So, we obtain the estimate:

$$E\sup_{0\le s\le t} |\zeta_N(t)|^2 \le C_0 + C_1\int_0^T \zeta_N^*(t)\ dt + C_2\int_0^T |\zeta_N^*(t)|^2\ dt. \tag{2.1.47}$$

Given the boundedness of the integral

$$\int_0^T \zeta_N^*(t)\ dt \le C < +\infty, \quad N \to \infty, \tag{2.1.48}$$

we apply the Gronwall–Bellman lemma

$$E\sup_{0\le s\le t} |\zeta_N^*(t)|^2 \le C_{01} + C_2\int_0^t E|\zeta_N^*(s)|^2\ ds, \quad 0 \le t \le T, \tag{2.1.49}$$

which justifies the assertion of Lemma 2.1.3. □

2.2. Multivariate DMPs in discrete–continuous time

The diffusion approximation of multivariate DMPs in discrete–continuous time $k = [Nt]$, $t \in R_+ := [0, +\infty)$, is realized for normalized SE fluctuations:

$$\zeta_N^{(m)}(t) := \sqrt{N}[S_N^{(m)}(k) - \rho_m], \quad 0 \le m \le M, \quad k = [Nt], \quad t \ge 0. \tag{2.2.1}$$

The DMP connection with SE [2.2.1] means that

$$S_N^{(m)}(k) = \rho_m + \zeta_N^{(m)}(t)/\sqrt{N}, \quad 0 \le m \le M, \quad k = [Nt], \quad t \ge 0. \tag{2.2.2}$$

The global balance condition for frequency SEs:

$$\sum_{m=0}^{M} S_N^{(m)}(k) \equiv 1 \quad \forall k \geq 0, \qquad [2.2.3]$$

generates the global balance condition for the DMP:

$$\sum_{m=0}^{M} \zeta_N^{(m)}(t) \equiv 0 \quad \forall t \geq 0. \qquad [2.2.4]$$

In asymptotic analysis of the DMP [2.2.1] by $N \to \infty$, the approach given in section 1.3 for multivariate statistical experiments (MSEs) is used.

2.2.1. *Evolutionary DMPs in discrete–continuous time*

The models of MSEs are defined by two components: predictable and stochastic.

The predictable component is given by conditional mathematical expectations (section 1.3):

$$V_0^{(m)}(P(k)) := E[\Delta S_N^{(m)}(k+1) \mid S_N(k) = P(k)], \quad k \geq 0. \qquad [2.2.5]$$

The RFI dependence on the vector parameter $p = p_m \ (0 \leq m \leq M)$ is postulated in Proposition 1.3.2:

$$V_0^{(m)}(p) := V_0 \pi p_m \left[\sum_{n=0}^{M} p_n^2/\rho_n - p_m/\rho_m \right], \quad 0 \leq m \leq M. \qquad [2.2.6]$$

Using the parameters $\pi_m = \pi/\rho_m$, $0 \leq m \leq M$, $\pi := \prod_{m=0}^{M} \rho_m$, the canonical RFI [1.3.9] transforms into a fluctuation representation

$$V_0^{(m)}(p, \widehat{p}) := V_0 p_m \left[\sum_{n=0}^{M} p_n \pi_n \widehat{p}_n - \pi_m \widehat{p}_m \right], \quad 0 \leq m \leq M. \qquad [2.2.7]$$

PROPOSITION 2.2.1.– *(Basic assumption 1)*. The frequency EPs

$$P_m(k+1) := E[\Delta S_N^{(m)}(k+1) \mid S_N(k) = P(k)],$$
$$0 \leq m \leq M, \ k \geq 0, \qquad [2.2.8]$$

are determined by the DEE solutions with the nonlinear RFI [2.2.7]:

$$\Delta \widehat{P}_m(k+1) = -V_0^{(m)}(P(k)), \widehat{P}(k)), \quad 0 \leq m \leq M, \quad k \geq 0. \qquad [2.2.9]$$

2.2.2. SDEs for the DMP in discrete–continuous time

Consider the convergence to the equilibrium state, defined by the equilibrium of the RFI:

$$\rho := (\rho_m,\ 0 \le m \le M), \quad \sum_{m=0}^{M} \rho_m = 1. \qquad [2.2.10]$$

DEFINITION 2.2.1.– The stochastic components of the DMPs are determined by the following martingale differences:

$$\Delta\mu_N^{(m)}(k+1) := \Delta S_N^{(m)}(k+1) - E[\Delta S_N^{(m)}(k+1) \mid S_N(k)], \quad k \ge 0. \qquad [2.2.11]$$

The conditional mathematical expectations in [2.2.5] are determined by the RFI [2.2.7] with normalized fluctuation EPs:

$$V_0^{(m)}(S_N(k), \zeta_N(t)) := V_0 \cdot \left[S_N^{(m)}(k) \sum_{n=0}^{M} \pi_n S_N^{(n)}(k) \zeta_N^{(n)}(t) - \pi_m S_N^{(m)}(k) \zeta_N^{(m)}(t) \right], \quad 0 \le m \le M, \quad k \ge 0. \qquad [2.2.12]$$

PROPOSITION 2.2.2.– (*Basic assumption 2*). The frequency DMP [2.2.1] is determined by the solution of the following SDE

$$\Delta\zeta_N^{(n)}(t+\Delta) = \frac{1}{N} V_0^{(m)}(S_N(k), \zeta_N(t)) + \frac{1}{\sqrt{N}} \sigma_m(S_N(k)) \Delta\mu_N^{(n)}(t+\Delta), \quad 0 \le m \le M, \quad k \ge 0. \qquad [2.2.13]$$

The stochastic component is characterized by conditional quadratic characteristic:

$$\sigma_m^2(S_N(k)) := E[(\Delta\mu_N^{(m)}(t+\Delta))^2 \mid S_N(k)] = V^{(m)}(S_N^{(m)}(k))[1 - V^{(m)}(S_N^{(m)}(k))], \quad 0 \le m \le M, \quad k \ge 0. \qquad [2.2.14]$$

LEMMA 2.2.1.– *The predictable component as well as the quadratic characteristic of the stochastic component of the SDE* [2.2.13] *converge in probability, by* $N \to \infty$*:*

$$V_0^{(m)}(S_N(k), \zeta_N(t)) \xrightarrow{P} -V_0\pi\zeta_N^{(m)}(t), \quad k \geq 0,$$

$$\sigma_m^2(S_N(k)) \xrightarrow{P} \rho_m(1-\rho_m), \quad 0 \leq m \leq M, \quad k \geq 0. \qquad [2.2.15]$$

Proof of Lemma [2.2.1].– There is a convergence in probability of the predictable component

$$V_0^{(m)}(S_N(k), \zeta_N(t)) \xrightarrow{P} -V_0\pi\zeta_N^{(m)}(t), \ N \to \infty, \ 0 \leq m \leq M, \ k \geq 0, \qquad [2.2.16]$$

and also the convergence in probability of the quadratic characteristic (see [2.2.14])

$$\sigma_m^2(S_N(k)) = \sigma_m^2\left(\rho + \frac{1}{\sqrt{N}}\zeta_N(t)\right) \xrightarrow{P} \rho_m(1-\rho_m), \quad N \to \infty. \qquad [2.2.17]$$

By virtue of [2.2.12], the predictable component has the expression:

$$V_0\left[\rho_m \sum_{n=0}^{M} \rho_n\pi_n\zeta_N^{(n)}(t) - \pi_m\zeta_N^{(m)}(t)\right]. \qquad [2.2.18]$$

Considering the identity $\rho_n\pi_n = \pi$, as well as the global balance condition [2.2.4], the nonlinear term in [2.2.18] disappears:

$$V_0 \sum_{n=0}^{M} \rho_n\pi_n\zeta_N^{(n)}(t) = V_0\pi \sum_{n=0}^{M} \zeta_N^{(n)}(t) \equiv 0. \qquad [2.2.19]$$

So there remains the following term:

$$-V_0\rho_m\pi_m\zeta_N^{(m)}(t) = -V_0\pi\zeta_N^{(m)}(t).$$

Thus, the nonlinear predictable component of the SDE [2.2.18] converges, in probability, to a linear component [2.2.15], which is identical for all m, where $0 \leq m \leq M$.

The convergence, in probability, of the stochastic component of the SDE [2.2.15] is obvious because of [2.2.17]. The proof of Lemma 2.2.1 is complete. □

An important consequence of Lemma 2.2.1 is a linear approximation of a nonlinear DMP formulated as follows.

PROPOSITION 2.2.3.– (*Basic assumption 3*). The frequency nonlinear DMPs are approximated by the solutions of the linear SDEs

$$\Delta\zeta_N^{(m)}(t+\Delta) = -\frac{1}{N}V_0\pi\zeta_N^{(m)}(t) + \frac{1}{\sqrt{N}}\breve{\sigma}_m\Delta\mu_N^{(m)}(t+\Delta),$$
$$0 \le m \le M, \quad \breve{\sigma}_m^2 = \rho_m(1-\rho_m). \qquad [2.2.20]$$

2.2.3. *Diffusion approximation of DMPs in discrete–continuous time*

We study approximation, using an Ornstein–Uhlenbeck continuous process, of multivariate nonlinear DMPs, determined by SDE solutions [2.2.13] in a discrete–continuous time $k = [Nt]$, $t \in R_+ := [0, +\infty)$, by $N \to \infty$, hence also by $k \to \infty$.

In this case, we substantially use the linear DMP [2.2.20], which approximates the nonlinear DMP [2.2.13].

THEOREM 2.2.1.– *By the condition of convergence, in probability, of initial conditions*

$$\zeta_N^{(m)}(0) \xrightarrow{P} \zeta^{(m)}(0)\ , \ 0 \le m \le M\ , \ N \to \infty, \qquad [2.2.21]$$

the convergence in distribution of the linear solutions of the SDE [2.2.20] *takes place*

$$\zeta_N^{(m)}(t) \xrightarrow{D} \zeta^{(m)}(t)\ , \ 0 \le m \le M\ , \ 0 \le t \le T\ , \ N \to \infty. \qquad [2.2.22]$$

The limit diffusion process with evolution $\zeta^{(m)}(t)$, $0 \le t \le T$, *is set by a generator*

$$L^{(m)}\varphi(c) = V_0\pi c\varphi'(c) + \frac{1}{2}\breve{\sigma}_m^2\varphi''(c), \ \ \varphi(c) \in C_0^\infty(R) \qquad [2.2.23]$$

on the class $C_0^\infty(R)$ *of real-valued functions, bounded together with their derivatives and having bounded supports.*
The parameters of the generator satisfy

$$V_0 > 0, \quad \pi = \prod_{m=0}^{M} \rho_m, \quad \breve{\sigma}_m^2 = \rho_m(1-\rho_m), \quad 0 \le m \le M.$$

Conclusion 2.2.1. The limit diffusion process with evolution is given by the differential stochastic equation

$$d\zeta^{(m)}(t) = V_0\pi\zeta^{(m)}(t)\,dt + \breve{\sigma}_m\,dW_m(t)\,,\ 0 \le m \le M, \quad t \ge 0. \quad [2.2.24]$$

The evolutional component is identical for all m, where $0 \le m \le M$.

The stochastic components are defined as the standard process of Brownian motion with dispersion $\breve{\sigma}_m^2 = \rho_m(1-\rho_m)$, $0 \le m \le M$:

$$E\,dW_m(t) = 0\,,\ E(dW_m(t))^2 = dt\,,\ 0 \le m \le M.$$

Proof of Theorem 2.2.1.– We use the operator and martingale characterization of the DMPs, which is determined by the SDE solution [2.2.11]–[2.2.14], taking into account the asymptotic properties of the parameters.

The generator of the DMP $\zeta(t)$, $t \ge 0$, is the following (see section 2.1):

$$\mathbb{L}_N\varphi(c) = NE[\varphi(c + \Delta\zeta_N(t+\Delta)) - \varphi(c)\,|\,\zeta_N(t) = c], \qquad [2.2.25]$$

acting on the class of functions $\varphi(c) \in C_0^\infty(R)$.

The martingale characterization of the DMPs is due to the ratio:

$$\int_0^{[Nt]/N} \mathbb{L}_N\varphi(\zeta_N(s))\,ds$$

$$= \sum_{k=0}^{t_N-1} E[\varphi(c + \Delta\zeta_N(t_k+\Delta)) - \varphi(c)\,|\,\zeta_N(t_k) = c]$$

$$= E\sum_{k=0}^{t_N-1}[\varphi(\zeta_N(t_{k+1})) - \varphi(\zeta_N(t_k))] = E[\varphi(\zeta_N(t_N)) - \varphi(\zeta_N(0))]. \qquad [2.2.26]$$

Here by definition

$$t_N := [Nt]/N = k_N/N \to T, \quad N \to \infty.$$

The key point in the proof of Theorem 2.2.1 is formulated in the following lemma.

LEMMA 2.2.2.– *The following asymptotic representation of the generator* [2.2.25] *takes place:*

$$\mathbb{L}_N\varphi(c) = \mathbb{L}\varphi(c) + \mathbb{R}_N\varphi(c), \quad \varphi(c) \in C_0^\infty(R), \qquad [2.2.27]$$

on the class of functions $\varphi(c) \in C_0^\infty(R)$*, with a negligible term:*

$$\mathbb{R}_N\varphi(c) \to 0, \quad N \to \infty, \quad \varphi(c) \in C_0^\infty(R). \qquad [2.2.28]$$

The limit diffusion process is given by the generator [2.2.23].

Proof of Lemma 2.2.2.– Applying the Taylor expansion up to the third order:

$$\mathbb{L}_N\varphi(c) = NE[\Delta\zeta_N(t+\Delta) \mid \zeta_N(t) = c]\varphi'(c) \\ + \frac{1}{2}NE[(\Delta\zeta_N(t+\Delta))^2|\zeta_N(t) = c]\varphi''(c)+\mathbb{R}_N\varphi(c), \qquad [2.2.29]$$

the first two moments of DMP increments are calculated, taking into account [2.2.11]–[2.2.14]:

$$NE[\Delta\zeta_N^{(m)}(t+\Delta) \mid \zeta_N(t) = c] = V_0^{(m)}(\rho + c/\sqrt{N}), \\ NE[(\Delta\zeta_N^{(m)}(t+\Delta))^2 \mid \zeta_N^{(m)}(t) = c] = \sigma_m^2(\rho + c/\sqrt{N}). \qquad [2.2.30]$$

Next, the asymptotic properties of predictable and stochastic components are used:

$$V_0^{(m)}(\rho + c/\sqrt{N}) \to -V_0\pi c, \quad N \to \infty, \qquad [2.2.31]$$

$$\sigma_m^2(\rho + c/\sqrt{N}) \to \sigma_m^2, \quad N \to \infty, \\ \sigma_m^2 := \rho_m(1-\rho_m), \quad 0 \le m \le M. \qquad [2.2.32]$$

Combining the asymptotic representations [2.2.29]–[2.2.32], we obtain the statement [2.2.27] of Lemma 2.2.2. □

The martingale characterization of the DMP [2.2.26] implies that

$$\mu_N^{(m)}(t) = \int_0^{[Nt]/N} \mathbb{L}_N\varphi(\zeta_N^{(m)}(s))\,\mathrm{d}s - \varphi(\zeta_N(t_N)) + \varphi(\zeta_N(0)), \quad t \ge 0, \qquad [2.2.33]$$

together with the asymptotic representation of the DMP generator [2.2.27] and [2.2.28], which ensures the fulfillment of the statement of Theorem 2.2.1. □

2.3. A DMP in an MRE

Consider a DMP in an MRE which drift parameters depending on the states of ergodic Markov chains (MCs) that determines the change of states of an MRE (Koroliouk 2015b).

The MRE is considered in two assumptions:

1. *Discrete MRE*, given by an MC x_k, $k \geq 0$, in a measurable state space $(E, \mathcal{E})$ with transition probabilities

$$P(x, B) = \mathcal{P}\{x_{k+1} \in B \,|\, x_k = x\}, \quad x \in E, \quad B \in \mathcal{E}. \tag{2.3.1}$$

2. *Continuous MRE*, given by an MP $x(t)$, $t \geq 0$, in a measurable state space $(E, \mathcal{E})$ with transition probabilities [2.3.1] of the embedded MC $x_k = x(\tau_k)$, $k \geq 0$, and by distributions of the sojourn times θ_{k+1}, $k \geq 0$:

$$\mathcal{P}\{\theta_{k+1} \geq t \,|\, x_k = x\} = \exp[-q(x)t], \quad t \geq 0. \tag{2.3.2}$$

So, the renewal moments are determined as

$$\tau_{k+1} = \tau_k + \theta_{k+1}, \quad k \geq 0. \tag{2.3.3}$$

2.3.1. *Discrete and continuous MRE*

The diffusion approximation for DMPs in the MRE are investigated using the method of singular perturbation (Korolyuk and Limnios 2005, Ch. 5). By tradition, the generator of two-component MP is given by singular perturbation with small series parameter $\varepsilon > 0$ $(\varepsilon \to 0)$. Next, we use the small parameter $\varepsilon^2 = 1/N$, or elsewhere $N = \varepsilon^{-2}$.

The transition to the small parameter ε clearly uses the singular perturbation technique.

Basic assumption 1. The SEs in a discrete MRE with discrete–continuous time $t = k\varepsilon^2$, $k \geq 0$, is given by the solution of the SDEs

$$\Delta\zeta_\varepsilon(t) = -\varepsilon^2 V(x_k)\zeta_\varepsilon(t) + \varepsilon\,\sigma(x_k)\Delta\mu_\varepsilon(t), \quad t = k\varepsilon^2, \quad k \geq 0. \tag{2.3.4}$$

The parameters $V(x) > 0$ and $\sigma(x)$, $x \in E$, are bounded real-valued functions of the sates $x \in E$ of an MC x_k, $k \geq 0$, which determines the MRE.

Here by definition

$$\Delta\zeta_\varepsilon(t) := \zeta_\varepsilon(t + \varepsilon^2) - \zeta_\varepsilon(t), \quad t \geq 0.$$

The stochastic component $\Delta\mu_\varepsilon(t)$, $t \geq 0$, is given by the martingale differences, characterized by the first two moments:

$$E[\Delta\mu_\varepsilon(t)] = 0, \quad E[[\Delta\mu_\varepsilon(t)]^2 \mid \zeta_\varepsilon(t)] = 1. \qquad [2.3.5]$$

The DMP in the MRE, given by the solutions of the SDEs [2.3.4] and [2.3.5], admits an approximation in a series scheme, with the series parameter $\varepsilon \to 0$, by a normal Ornstein–Uhlenbeck process type with continuous time under additional conditions.

THEOREM 2.3.1.– *Let the following conditions be fulfilled:*

C1: *The MC with transition probabilities* [2.3.1] *is uniformly ergodic with the stationary distribution* $\rho(B)$, $B \in \mathcal{E}$:

$$\rho(B) = \int_E \rho(dx) P(x, B), \quad \rho(\mathcal{E}) = 1. \qquad [2.3.6]$$

C2: *The martingale differences satisfy the conditions of the central limit theorem (Borovskikh and Korolyuk 1997).*

C3: *There is the convergence and uniform boundedness of initial conditions:*

$$\zeta_\varepsilon(0) \xrightarrow{P} \zeta(0), \quad \varepsilon \to 0, \quad E|\zeta_\varepsilon(0)| \leq C < +\infty. \qquad [2.3.7]$$

Then a convergence in distribution of the DMP takes place:

$$\zeta_\varepsilon(t) \xrightarrow{D} \zeta(t), \quad \varepsilon \to 0, \quad t \in [0, T]. \qquad [2.3.8]$$

The limit process $\zeta(t)$, $t \geq 0$, *is set by the generator*

$$\mathbb{L}\varphi(s) = -\widehat{V} s\, \varphi'(s) + 1/2\, \widehat{\sigma}^2 \varphi''(s). \qquad [2.3.9]$$

The drift parameters $\widehat{V}$ and dispersions $\widehat{\sigma}^2$ are calculated by the averaging formulas:

$$\widehat{V} = \int_E \rho(\mathrm{d}x)V(x), \quad \widehat{\sigma}^2 = \int_E \rho(\mathrm{d}x)\sigma^2(x). \qquad [2.3.10]$$

REMARK 2.3.1.– The limit process $\zeta(t)$, $t \geq 0$, with generators [2.3.9] and [2.3.10] is a normal Ornstein–Uhlenbeck diffusion process, determined by the stochastic differential equation:

$$\mathrm{d}\zeta(t) = -\widehat{V}\zeta(t)\,\mathrm{d}t + \widehat{\sigma}\,\mathrm{d}W(t), \quad t \geq 0. \qquad [2.3.11]$$

Here $\mathrm{d}W(t)$, $t \geq 0$, is a standard process of Brownian motion with parameters

$$E[\mathrm{d}W(t)] = 0, \quad E[\mathrm{d}W(t)]^2 = dt.$$

In contrast to the DMP in a discrete MRE, in which the DMP dynamics is determined at each (discrete–continuous) time instant by the solution of the SDE [2.3.4], the DMP dynamics in a continuous MRE is determined at the instants of increment of the MP.

An extension of the DMP in a continuous MRE is assumed, according to the scheme:

$$\zeta_\varepsilon(t) = \zeta_\varepsilon(t_k), \quad t_k = k\varepsilon^2 \leq t < t_{k+1} = t_k + \varepsilon^2, \quad k \geq 0.$$

Basic assumption 2. The DMP in a continuous MRE is given by the solutions of the SDE

$$\Delta\zeta_\varepsilon(\tau_{k+1}) = -\varepsilon^2 V(x_k)\zeta_\varepsilon(\tau_k) + \varepsilon\,\sigma(x_k)\Delta\mu_\varepsilon(\tau_{k+1}). \qquad [2.3.12]$$

The stochastic component $\Delta\mu_\varepsilon(\tau_k)$, $k \geq 0$, is characterized by the first two moments:

$$E[\Delta\mu_\varepsilon(\tau_{k+1}) \mid \zeta_\varepsilon(\tau_k)] = 0, \quad E[[\Delta\mu_\varepsilon(\tau_{k+1})]^2 \mid \zeta_\varepsilon(\tau_k)] = 1. \quad [2.3.13]$$

Consider the accompanying functions, relative to the MPs $x(t)$, $t \geq 0$:

$$\nu(t) := \max\{k : \tau_k \leq t\}, \quad t \geq 0,$$

$$\tau(t) := \tau_{\nu(t)}, \quad \theta(t) := \tau_{\nu(t)+1} - \tau(t).$$

The DMP in discrete–continuous time $t = k\varepsilon^2$, $k \geq 0$, is given by the sum of increments

$$\zeta_\varepsilon(t) = \zeta_\varepsilon(0) + \sum_{k=1}^{\nu(t/\varepsilon^2)} \Delta\zeta_\varepsilon(\tau_k), \quad t = k\varepsilon^2, \quad k \geq 0. \qquad [2.3.14]$$

THEOREM 2.3.2.– *Let the following conditions be fulfilled:*

C1: *The MP $x(t)$, $t \geq 0$, be uniformly ergodic with stationary distribution $\pi(B)$, $B \in \mathcal{E}$:*

$$\pi(\mathrm{d}x)q(x) = q\rho(\mathrm{d}x), \quad q = \int_E \pi(\mathrm{d}x)q(x). \qquad [2.3.15]$$

C2: *The martingale differences $\Delta\mu_\varepsilon(\tau_k)$, $k \geq 1$, satisfy the central limit theorem conditions (Borovskikh and Korolyuk 1997);*

C3: *There is a convergence of the initial conditions:*

$$\zeta_\varepsilon(0) \xrightarrow{P} \zeta(0), \quad \varepsilon \to 0, \quad E|\zeta_\varepsilon(0)| \leq C < +\infty. \qquad [2.3.16]$$

Then there is the convergence in distribution of the DMPs:

$$\zeta_\varepsilon(t) \xrightarrow{D} \zeta(t), \ \ \varepsilon \to 0, \quad t \in [0, T]. \qquad [2.3.17]$$

The limit process $\zeta(t)$, $t \geq 0$, is set by the generator

$$\mathbb{L}\varphi(s) = -\widehat{V} s\, \varphi'(s) + 1/2\, \widehat{\sigma}^2 \varphi''(s). \qquad [2.3.18]$$

The drift parameter $\widehat{V}$ and dispersion $\widehat{\sigma}^2$ are calculated by the corresponding averages:

$$\widehat{V} = q \int_E \rho(\mathrm{d}x)V(x), \quad \widehat{\sigma}^2 = q \int_E \rho(\mathrm{d}x)\sigma^2(x). \qquad [2.3.19]$$

REMARK 2.3.2.– The limit process $\zeta(t)$, $t \geq 0$, with generators [2.3.18] and [2.3.19] is a normal diffusion process of Ornstein–Uhlenbeck type, determined by the stochastic differential equation [2.3.11].

2.3.2. *Proof of limit theorems 2.3.1 and 2.3.2*

The basic idea of proving limit theorems 2.3.1 and 2.3.2 for the MPs is the application of limit theorems for random processes, in particular, for random evolutions, based on the convergence of generators for the corresponding MPs (Skorokhod *et al.* 2002; Korolyuk and Limnios 2005; Korolyuk 1999; Ethier and Kurtz 1986). The convergence of generators, on a sufficiently rich class of real-valued functions, ensures the convergence of finite-dimensional distributions (Skorokhod *et al.* 2002).

The presence of random Markovian environment necessitates the technique of *singular perturbation problem* for reducible invertible operator that determines a uniformly ergodic MC (or MP) (Korolyuk and Limnios 2005).

First of all, the characterization of the extended MP with additional component, which sets the MRE, is used.

LEMMA 2.3.1.– ***A:*** *For a discrete MRE, the following two-component MC*

$$\left(\zeta^d_\varepsilon(t;x),\, x_\varepsilon(t) := x(t/\varepsilon^2)\right), \quad t = k\varepsilon^2, \quad k \geq 0, \qquad [2.3.20]$$

is determined by the generator

$$\mathcal{L}^d_\varepsilon(x)\varphi(c,x) = \varepsilon^{-2}\left[\Gamma^d_\varepsilon(x)\mathbb{P} - \mathbb{I}\right]\varphi(c,x), \qquad [2.3.21]$$

where the operators $\Gamma^d_\varepsilon(x)$ *and* $\mathbb{P}$ *are transition probabilities of the MC component* [2.3.20], *defined by the following formule:*

$$\Gamma^d_\varepsilon(x)\varphi(c) := E\left[\varphi\left(c + \Delta\zeta^d_\varepsilon(t;x)\right)\,\Big|\,\zeta^d_\varepsilon(t) = c\,,\, x_\varepsilon(t) = x\right], \quad t \geq 0.$$

$$[2.3.22]$$

$$\mathbb{P}\varphi(x) := \int_E P(x,dy)\varphi(y). \qquad [2.3.23]$$

B: *For a continuous MRE, the two-component MP*

$$(\zeta^c_\varepsilon(t;x),\, x_\varepsilon(t) = x(k)), \quad k = t/\varepsilon^2, \quad t \geq 0, \qquad [2.3.24]$$

is determined by the generator

$$\mathcal{L}^c_\varepsilon(x)\varphi(c,x) = q(x)\mathcal{L}^d_\varepsilon(x)\varphi(c,x). \qquad [2.3.25]$$

Proof of Lemma 2.3.1.– The assertion A follows from the definition of the MC generator in discrete–continuous time with elementary interval $\Delta = \varepsilon^2$ (see Nevelsona and Has'minskii 1976, Ch. 2, section 1).

The assertion B follows from the well-known presentation for the generator of the MP with continuous time, taking into account the transition probabilities of the embedded MC. □

A significant step in the proof of Theorems 2.3.1 and 2.3.2 is realized in the following lemma.

LEMMA 2.3.2.– *The generator* [2.3.21] *of two-component MC* [2.3.20] *admits an asymptotic expansion on the real-valued test functions* $\varphi(c,x)$, *with three bounded derivatives by c:* $\varphi(c,\cdot) \in C^3(R)$*:*

$$\mathcal{L}_\varepsilon^d(x)\varphi(c,x) = \left[\varepsilon^{-2}\mathbb{Q} + \mathbb{L}^0(x)\mathbb{P}\right]\varphi(c,x) + \mathbb{R}_\varepsilon(x)\varphi(c,x). \quad [2.3.26]$$

Here by definition

$$\mathbb{Q}\varphi(\cdot,x) := [\mathbb{P} - \mathbb{I}]\varphi(\cdot,x), \quad [2.3.27]$$

$$\mathbb{L}^0(x)\varphi(c) := -V(x)c\,\varphi'(c) + 1/2\,\sigma^2(x)\varphi''(c). \quad [2.3.28]$$

The residual term

$$|\mathbb{R}_\varepsilon(x)\varphi(c,x)| \to 0,\ \varepsilon \to 0,\ \ \varphi(c,x) \in C^3(R). \quad [2.3.29]$$

Proof of Lemma 2.3.2 is based on the following transformation of generator [2.3.21]:

$$\mathcal{L}_\varepsilon^d(x)\varphi(c,x) = \varepsilon^{-2}[\mathbb{Q} + \mathbb{L}_\varepsilon^d(x)\mathbb{P}]\varphi(c,x). \quad [2.3.30]$$

Here by definition

$$\mathbb{L}_\varepsilon^d(x)\varphi(c,\cdot) := \varepsilon^{-2}[\Gamma_\varepsilon^d(x) - \mathbb{I}]\varphi(c,\cdot). \quad [2.3.31]$$

Then we use the Taylor formula for test functions $\varphi(c)$ up to third order inclusively:

$$\mathbb{L}_\varepsilon^d(x)\varphi(c) = \mathbb{L}^0(x)\varphi(c) + \mathbb{R}_\varepsilon(x)\varphi(c) \quad [2.3.32]$$

with a residual term that satisfies the condition [2.3.29].

The generator representation [2.3.26] conducts to a singular perturbation problem for the truncated operator (see Korolyuk and Limnios 2005, Ch. 5):

$$\mathcal{L}^0_\varepsilon \varphi(c,x) := [\varepsilon^{-2}\mathbb{Q} + \mathbb{L}^0(x)\mathbb{P}]\varphi(c,x). \qquad [2.3.33]$$

LEMMA 2.3.3.– *On the perturbed test functions*

$$\varphi_\varepsilon(c,x) = \varphi(c) + \varepsilon^2\varphi_1(c,x), \qquad [2.3.34]$$

the solution of the singular perturbation problem (Korolyuk and Limnios 2005, Proposition 5.1) for the truncated operator [2.3.33] *is determined by the relation:*

$$\mathcal{L}^0_\varepsilon(x)\varphi_\varepsilon(c,x) = \mathbb{L}^0\varphi(c) + \mathbb{R}_\varepsilon(x)\,\varphi(c). \qquad [2.3.35]$$

Here the limit operator

$$\mathbb{L}^0\varphi(c) = -\widehat{V}c\varphi'(c) + \frac{1}{2}\widehat{\sigma}^2\varphi''(c). \qquad [2.3.36]$$

The residual term $\mathbb{R}_N(x)$ *satisfies the condition* [2.3.29].

Proof of Lemma 2.3.3.– On the perturbed test functions

$$\varphi(c) + \varepsilon^2\varphi_1(c,x),$$

we can calculate the generator

$$\begin{aligned}\mathcal{L}^0_\varepsilon(x)\varphi_\varepsilon(c,x) &= [\varepsilon^{-2}\mathbb{Q} + \mathbb{L}^0(x)\mathbb{P}][\varphi(c) + \varepsilon^2\varphi_1(c,x)]\\ &= \varepsilon^{-2}\mathbb{Q}\varphi(c) + [\mathbb{Q}\varphi_1(c,x) + \mathbb{L}^0(x)\mathbb{P}\varphi(c)] + \varepsilon^2\mathbb{L}^0(x)\mathbb{P}\varphi_1(c,x). \qquad [2.3.37]\end{aligned}$$

Taking into account that $\mathbb{Q}\varphi(c) = 0$, as well as the representation

$$\mathbb{Q}\varphi(x) := [\mathbb{P} - \mathbb{I}]\varphi(x),$$

the first singular term is a the generating operator of an ergodic MC (see C1 in Theorem 2.3.1), using the solvability condition of the equation

$$\mathbb{Q}\varphi_1(c,x) + \mathbb{L}^0(x)\mathbb{P}\varphi(c) = \mathbb{L}^0\varphi(c), \qquad [2.3.38]$$

which ensures the existence of a test function $\varphi_1(c,x)$ (see Korolyuk and Limnios 2005, Proposition 5.1).

The averaged operator $\mathbb{L}^0$ is determined by

$$\mathbb{L}^0 = \Pi\mathbb{L}^0(x)\mathbb{P}\Pi = \Pi\mathbb{L}^0(x)\Pi. \qquad [2.3.39]$$

The calculation, by using formula [2.3.39], defines the limit operator [2.3.9] and [2.3.10]. □

At the end of the proof of Theorem 2.3.1, we apply the model limit theorem (Korolyuk and Limnios 2005, Ch. 6, section 3), which ensures the convergence of finite-dimensional distributions.

Proof of Theorem 2.3.2 is completed with calculations of solvability conditions for the generator $\mathbb{Q}^c = q(x)\mathbb{Q}$ of equation

$$\mathbb{Q}^c\varphi_1(c,x) + q(x)\mathbb{L}^0(x)\mathbb{P}\varphi(c) = \mathbb{L}^c\varphi(c), \qquad [2.3.40]$$

where

$$\mathbb{L}^c = \Pi q(x)\mathbb{L}^0(x)\mathbb{P}\Pi. \qquad [2.3.41]$$

Relation [2.3.41] with the projector

$$\Pi\varphi(x) := \int_E \pi(dx)\varphi(x), \qquad [2.3.42]$$

gives the limit operator [2.3.18] with parameters [2.3.19].

REMARK 2.3.3.– The existence of perturbation $\varphi_1(c,x)$, which satisfies equation [2.3.38], provides the condition for solvability

$$\Pi[\mathbb{L} - \mathbb{L}^0(x)\mathbb{P}]\varphi(c) = 0.$$

Then the solution of equation [2.3.38] exists:

$$\varphi_1(s,x) = \mathbb{R}_0^d\widetilde{\mathbb{L}}(x)\varphi(s), \quad \widetilde{\mathbb{L}}(x) := \mathbb{L} - \mathbb{R}_0^d\mathbb{L}^0(x),$$

where $\mathbb{R}_0^d$ is a generalized inverse operator

$$\mathbb{Q}^d\mathbb{R}_0^d = \mathbb{R}_0^d\mathbb{Q}^d = \Pi - \mathbb{I}.$$

In this case, the residual term in the expansion [2.3.37] has the form

$$\varepsilon^2\mathbb{C}^0(x)\mathbb{P}\mathbb{R}_0^d\widetilde{\mathbb{L}}(x)\varphi(c) \to 0, \quad \varepsilon \to 0.$$

2.4. The DMPs in a balanced MRE

The *DMPs with discrete–continuous time* $t = [k\varepsilon^2]$, $k \geq 0$, are given by the solutions of the SDEs with two components: predictable and stochastic (martingale differences) (see section 2.3). The RFIs, as well as the diffusion coefficient of the stochastic component, depend on the states of the embedded MC in homogeneous (in time), uniformly ergodic MP, which describes the states of the random environment. The balanced MRE indicates zero mean of the regression increments over the ergodic distribution of the MP, which corresponds to the random environment.

Here the approximation using an Ornstein–Uhlenbeck-type diffusion process in the series scheme is obtained, with a small parameter ε ($\varepsilon \to 0$). The drift and diffusion parameters are determined by averaging over the stationary distribution of the embedded MC, taking into account the balance condition.

2.4.1. *Basic assumptions*

The Markov SEs with discrete–continuous time are given by the solutions of the SDEs with the parameters of predictable and stochastic components, which depend on the states of an MRE.

The MRE is considered in two assumptions:

1. *Discrete MRE*, given by an ergodic MC x_k, $k \geq 0$, in a measurable state space $(E, \mathcal{E})$ with transition probabilities:

$$P(x, B) = \mathcal{P}\{x_{k+1} \in \mathcal{B} \,|\, x_k = x\}, \quad x \in E, \quad \mathcal{B} \in \mathcal{E}. \qquad [2.4.1]$$

2. *Continuous MRE*, given by ergodic MP $x(t)$, $t \geq 0$, in a measurable state space $(E, \mathcal{E})$ with transition probabilities [2.4.1] of an embedded MC $x_k = x(\tau_k^{\varepsilon})$, $k \geq 0$, and distributions of sojourn times in the states θ_{k+1}, $k \geq 0$:

$$\mathcal{P}\{\theta_{k+1} \geq t \,|\, x_k = x\} = \exp[-q(x)t], \quad t \geq 0. \qquad [2.4.2]$$

So, the renewal instant has the form

$$\tau_{k+1}^{\varepsilon} = \tau_k^{\varepsilon} + \varepsilon^2 \theta_{k+1}, \quad k \geq 0. \qquad [2.4.3]$$

The drift parameter $V(x)$, $x \in E$, can take both positive and negative values with the balance condition for a random environment:

$$\int_E \rho(dx) V(x) = 0. \qquad [2.4.4]$$

Consequently, in a balanced environment, the stationary averaged drift increment is zero. In this connection, the normalization of the components of the SDE for SEs is different.

Basic assumption A in a discrete MRE. The DMP in a balanced MRE [2.4.1] with discrete–continuous time $t = k\varepsilon^2$, $k \geq 0$, is given by the solution of the SDE (compare with [2.3.4]):

$$\Delta\zeta^\varepsilon(t) = \varepsilon\,[V(x_k)\zeta^\varepsilon(t) + \sigma(x_k)\Delta\mu^\varepsilon(t)], \quad k \geq 0, \qquad [2.4.5]$$

under additional conditions

$$E[\Delta\mu^\varepsilon(t)] = 0, \quad E[[\Delta\mu^\varepsilon(t)]^2 \,|\, \zeta^\varepsilon(t)] = 1, \quad t \geq 0. \qquad [2.4.6]$$

The drift parameters $V(x)$ and $\sigma(x)$, $x \in E$, are bounded real-valued functions on the states $x \in E$ of the MC x_k, $k \geq 0$, which determines the MRE.

Here by definition

$$\Delta\zeta^\varepsilon(t + \varepsilon^2) := \zeta^\varepsilon(t + \varepsilon^2) - \zeta^\varepsilon(t), \quad t = [k\varepsilon^2].$$

Basic assumption B in a continuous MRE. We define the extension of the DMPs in a continuous MRE as a stepwise process:

$$\zeta^\varepsilon(t) = \zeta^\varepsilon(\tau_k^\varepsilon), \quad \tau_k^\varepsilon \leq t < \tau_{k+1}^\varepsilon, \quad k \geq 0.$$

The DMPs in a continuous MRE are given by the solutions of the SDE

$$\Delta\zeta^\varepsilon(t) = \varepsilon\,[V(x_k)\zeta^\varepsilon(\tau_k^\varepsilon) + \sigma(x_k)\Delta\mu^\varepsilon(\tau_{k+1}^\varepsilon)], \quad k \geq 0, \qquad [2.4.7]$$

with the additional conditions:

$$E[\Delta\mu^\varepsilon(\tau_{k+1}^\varepsilon) \,|\, \zeta^\varepsilon(\tau_k^\varepsilon)] = 0, \quad E[[\Delta\mu^\varepsilon(\tau_{k+1}^\varepsilon)]^2 \,|\, \zeta^\varepsilon(\tau_k^\varepsilon)] = 1, \quad k \geq 0. \qquad [2.4.8]$$

The DMPs, defined by the solutions of the SDEs [2.4.5] and [2.4.6], and also [2.4.7] and [2.4.8], under additional conditions, admit approximation in the series scheme with the parameter $\varepsilon \to 0$, by a normal Ornstein–Uhlenbeck process with continuous time. At the same time, the algorithm for calculating the drift parameter and the dispersion of the stochastic component in the limit process is significantly altered.

THEOREM 2.4.1.– *Let the following conditions be fulfilled:*

C0: *Balance condition*

$$\int_E \rho(\mathrm{d}x)V(x) = 0. \qquad [2.4.9]$$

C1A: *MC* x_k, $k \geq 0$, *with transitional probabilities* [2.4.1] *is uniformly ergodic with a stationary distribution* $\rho(B)$, $B \in \mathcal{E}$*:*

$$\rho(B) = \int_E \rho(\mathrm{d}x)P(x, B), \quad \rho(E) = 1. \qquad [2.4.10]$$

C1B: *MC* $x(t)$, $t \geq 0$, *with transitional probabilities* [2.4.1] *is uniformly ergodic with a stationary distribution* $\pi(B)$, $B \in \mathcal{E}$*:*

$$\pi(\mathrm{d}x)q(x) = q\rho(\mathrm{d}x), \quad q = \int_E \pi(\mathrm{d}x)q(x). \qquad [2.4.11]$$

C2: *The martingale differences* $\Delta\mu^\varepsilon(t)$, $t \geq 0$, *satisfy the condition*

$$E[\Delta\mu^\varepsilon(t)]^3 < C < +\infty \quad \forall t \in [0, T].$$

C3: *There is a convergence of the initial conditions:*

$$\zeta^\varepsilon(0) \xrightarrow{P} \zeta(0), \quad \varepsilon \to 0, \quad E\,|\zeta(0)| \leq C < +\infty. \qquad [2.4.12]$$

Then a convergence in distribution of DMPs takes place:

$$\zeta^\varepsilon(t) \xrightarrow{D} \zeta(t), \quad \varepsilon \to 0, \quad t \in [0, T]. \qquad [2.4.13]$$

In this case:

A: *In a discrete MRE, the limit process* $\zeta(t)$, $t \geq 0$, *is set by the generator*

$$\mathbb{L}_d^0\varphi(c) = \widehat{V_0 s}\varphi'(c) + 1/2\widehat{\sigma_0^2}(c)\varphi''(c). \qquad [2.4.14]$$

The drift parameter $\widehat{V_0}$ *and dispersions* $\widehat{\sigma_0^2}$ *are calculated by the averaging formulae:*

$$\widehat{V_0} = \widehat{V_0^2} - \widehat{V^2}, \quad \widehat{\sigma_0^2}(c) = \widehat{\sigma^2} + (2\widehat{V_0^2} - \widehat{V^2})c^2, \qquad [2.4.15]$$

where

$$\widehat{V_0^2} := \int_E \rho(\mathrm{d}x)V(x)\mathbb{R}_0 V(x), \quad \widehat{V^2} := \int_E \rho(\mathrm{d}x)V^2(x),$$

$$\widehat{\sigma^2} := \int_E \rho(\mathrm{d}x)\sigma^2(x). \qquad [2.4.16]$$

The potential operator $\mathbb{R}_0$ *is reducibly invertible with respect to the MC generator* $\mathbb{Q} := \mathbb{P} - \mathbb{I}$ *(Korolyuk 1999):*

$$\mathbb{Q}\,\mathbb{R}_0 = \mathbb{R}_0\,\mathbb{Q} = \Pi - \mathbb{I}. \qquad [2.4.17]$$

The projector Π *on the subspace of zeros of the operator* $\mathbb{Q}$ *is determined as follows:*

$$\Pi\varphi(x) = \int_E \rho(\mathrm{d}x)\varphi(x). \qquad [2.4.18]$$

B: *In a continuous MRE, the limit process* $\zeta(t)$*,* $t \geq 0$*, is set by the generator*

$$\mathbb{L}_c^0\varphi(c) = q\,\mathbb{L}_d^0\varphi(c). \qquad [2.4.19]$$

REMARK 2.4.1.– The limit process $\zeta(t)$, $t \geq 0$, with the generator given by equations [2.4.14]–[2.4.16] or by relation [2.4.19] is a normal Ornstein–Uhlenbeck diffusion process, given by the following stochastic differential equation:

$$\mathrm{d}\zeta(t) = -\widehat{V_0}\zeta(t)\,\mathrm{d}t + \widehat{\sigma_0}\,\mathrm{d}W(t), \quad t \geq 0. \qquad [2.4.20]$$

Here $W(t)$, $t \geq 0$, is a standard process of Brownian motion with the parameters:

$$E[W(t)] = 0, \quad E[[W(t)]^2] = t.$$

2.4.2. *Proof of limit theorem 2.4.1*

The main idea of the proof of limit theorem 2.4.1 for MPs is to apply limit theorems for random processes, in particular, for random evolutions, based on the convergence of generators, defining the corresponding MPs (Skorokhod *et al.* 2002; Korolyuk 1999; Korolyuk and Limnios 2005). The convergence of finite-dimensional distributions is ensured by the convergence of generating operators (generators) on a sufficiently large class of real-valued test functions.

In the sequel, we use the class $C_b^3(R)$ of real-valued functions $\varphi(c)$ with finite support (different from 0 on finite intervals) and three times continuously differentiated with bounded derivatives. In addition, using the test functions $\varphi(c,x)$, $x \in \mathcal{E}$, which depend on the additional argument x, the condition $\varphi(c,\cdot) \in C_b^3(R)$ means the existence of a function $\varphi(c) \in C_b^3(R)$, which limits, on the top, the function $\varphi(c,x) : \ |\varphi(c,x)| \leq \varphi(c)$.

The presence of a random Markov environment necessitates the use of singular perturbation theory for the reducibly invertible operator, which determines a uniformly ergodic MC (or MP) (Korolyuk and Limnios 2005).

First of all, the characterization of an extended MP with the additional component, which sets the MRE, is used.

LEMMA 2.4.1.– **A**: *In the case of a discrete MRE, the two-component MC*

$$\{\zeta^\varepsilon(t),\ x^\varepsilon(t) := x(t/\varepsilon^2)\}, \quad t = k\varepsilon^2, \quad \varepsilon > 0 \qquad [2.4.21]$$

is set by the generator

$$\mathbb{L}_d^\varepsilon(x)\varphi(c,x) = \varepsilon^{-2}[\mathbb{C}^\varepsilon(x)\mathbb{P} - \mathbb{I}]\varphi(c,x), \qquad [2.4.22]$$

where the operators $\mathbb{P}$ *and* $\mathbb{C}^\varepsilon(x)$ *are defined as follows:*

$$\mathbb{P}\varphi(x) := \int_E P(x, \mathrm{d}y)\varphi(y), \qquad [2.4.23]$$

$$\mathbb{C}^\varepsilon(x)\varphi(c) := E[\varphi(c + \Delta\zeta^\varepsilon(t)) \,|\, \zeta^\varepsilon(t) = c,\ x^\varepsilon(t) = x], \quad t \geq 0. \qquad [2.4.24]$$

B: *In the case of a continuous MRE, the two-component MC*

$$\{\zeta^\varepsilon(t),\ x^\varepsilon(t) := x(t/\varepsilon^2)\}, \quad t = k\varepsilon^2, \quad k \geq 0, \qquad [2.4.25]$$

is set by the generator

$$\mathbb{L}_d^\varepsilon(x)\varphi(c,x) = \varepsilon^{-2}[\mathbb{C}^\varepsilon(x)\mathbb{P} - \mathbb{I}]\varphi(c,x). \qquad [2.4.26]$$

Proof of Lemma 2.4.1.– Statement ***A*** follows from the definition of the MC generator in discrete–continuous time with scaling ε^2 (see Nevelson and Has'minskii 1976, Ch. 2, section 1).

Statement ***B*** follows from the well-known definition of the generator of the MP with continuous time, taking into account the transition probabilities of the embedded MC. □

The essential step in the proof of Theorem 2.4.1 is realized in the following lemma.

LEMMA 2.4.2.– *The generator* [2.4.26] *of two-component MC* [2.4.21] *admits the following asymptotic decomposition on real-valued test functions* $\varphi(c,\cdot) \in C_b^3(R)$*:*

$$\mathbb{L}_d^\varepsilon(x)\varphi(c,x) = \Big[\varepsilon^{-2}\mathbb{Q} + \varepsilon^{-1}\mathbb{C}_1(x)\mathbb{P} + \mathbb{C}_2(x)\mathbb{P}\Big]\varphi(c,x) + \varepsilon\mathbb{R}^\varepsilon(x)\mathbb{P}\varphi(c,x). \qquad [2.4.27]$$

Here

$$\mathbb{Q}\varphi(\cdot,x) = [\mathbb{P} - \mathbb{I}]\varphi(\cdot,x), \qquad [2.4.28]$$

$$\mathbb{C}_1(x)\varphi(c,\cdot) = V(x)\,c\,\varphi'(c,\cdot),$$

$$\mathbb{C}_2(x)\varphi(c,\cdot) = 1/2[V^2(x)c^2 + \sigma^2(x)]\varphi''(c,\cdot), \qquad [2.4.29]$$

and the residual term with bounded $\mathbb{R}^\varepsilon(x)$ *(see* [2.4.34]*):*

$$\varepsilon\mathbb{R}^\varepsilon\varphi(c,x) \to 0, \quad \varepsilon \to 0, \quad \varphi(c,\cdot) \in C_b^3(R). \qquad [2.4.30]$$

Proof of Lemma 2.4.2.– Consider the following transformation of generator [2.4.26]:

$$\mathbb{L}_d^\varepsilon(x)\varphi(c,x) = \varepsilon^{-2}[\mathbb{Q} + \mathbb{C}_0^\varepsilon(x)\mathbb{P}]\varphi(c,x). \qquad [2.4.31]$$

Here by definition

$$\mathbb{C}_0^\varepsilon(x) := \mathbb{C}^\varepsilon(x) - \mathbb{I}. \qquad [2.4.32]$$

Further, taking into account the basic assumption [2.4.5], using the Taylor expansion of functions $\mathbb{C}_0^\varepsilon(x)\varphi(c)$ by the degrees of $\Delta\zeta_x^\varepsilon(t)$ in the neighborhood of the point $\zeta_x^\varepsilon(t) = c$, up to the third order, we obtain:

$$\begin{aligned}\mathbb{C}_0^\varepsilon(x)\varphi(c) &= E\varphi\Big(c + \Delta\zeta_x^\varepsilon(t+\varepsilon^2)\Big) - \varphi(c)\\ &= E\varphi\Big(c + \varepsilon\Big[V(x)\cdot c + \sigma(x)\Delta\mu^\varepsilon(t+\varepsilon^2)\Big]\Big) - \varphi(c)\\ &= \varepsilon V(x)c\varphi'(c) + \frac{1}{2}\varepsilon^2\Big[V^2(x)c^2 + \sigma^2(x)\Big]\varphi''(c) + \varepsilon^3\mathbb{R}^\varepsilon(x)\varphi(c),\end{aligned} \quad [2.4.33]$$

where

$$\begin{aligned}\mathbb{R}^\varepsilon(x)\varphi(c) &= 1/6E\Big[(\Delta\zeta_x^\varepsilon)^3\,\varphi'''(c)\Big]\\ &= 1/6\Big(V^3(x)c^3 + 3V(x)c\sigma^2(x) + \sigma^3(x)E[\Delta\mu^\varepsilon(x)]^3\Big)E\varphi'''(c+\theta\Delta\zeta_x^\varepsilon(t)).\end{aligned} \quad [2.4.34]$$

The residual term in [2.4.34], considering the basic assumption **A** and the condition **C2** of Theorem 2.4.1, is bounded on the class of functions $C_b^3(R)$. Consequently, condition [2.4.30] is fulfilled.

The next stage of the proof of Theorem 2.4.1 is the use of the singular perturbation technique for the truncated operator (see Korolyuk and Limnios 2005, Ch. 5):

$$\overline{\mathbb{L}_d^\varepsilon}(x)\varphi(c,x) = [\varepsilon^{-2}\mathbb{Q} + \varepsilon^{-1}\mathbb{C}_1(x)\mathbb{P} + \mathbb{C}_2(x)\mathbb{P}]\varphi(c,x). \quad [2.4.35]$$

LEMMA 2.4.3.– *The solution of the singular perturbation problem for the truncated operator* [2.4.35] *on perturbed test functions*

$$\varphi^\varepsilon(c,x) = \varphi(c) + \varepsilon\varphi_1(c,x) + \varepsilon^2\varphi_2(c,x), \quad [2.4.36]$$

is given by

$$\overline{\mathbb{L}_d^\varepsilon}(x)\varphi(c,x) = \mathbb{L}_d^0\varphi(c) + \mathbb{R}_2^\varepsilon(x)\varphi(c). \quad [2.4.37]$$

Here the limit operator $\mathbb{L}_d^0$ *is determined by the relations* [2.4.14]–[2.4.16], *and the residual term* $\mathbb{R}_2^\varepsilon(x)$ *is negligible:*

$$\mathbb{R}_2^\varepsilon(x)\varphi(c) \to 0, \quad \varepsilon \to 0, \quad \varphi(c) \in C_b^3(R). \quad [2.4.38]$$

Proof of Lemma 2.4.3.– Consider the relationship

$$\begin{aligned}
&\mathbb{L}_d^{\varepsilon}(x)\varphi(c,x)\\
&\quad = \left[\varepsilon^{-2}\mathbb{Q} + \varepsilon^{-1}\mathbb{C}_1(x)\mathbb{P} + \mathbb{C}_2(x)\mathbb{P}\right][\varphi(c) + \varepsilon\varphi_1(c,x) + \varepsilon^2\varphi_2(c,x)]\\
&\quad = \varepsilon^{-2}\mathbb{Q}\varphi(c) + \varepsilon^{-1}[\mathbb{Q}\varphi_1(c,x) + \mathbb{C}_1(x)\varphi(c)]\\
&\qquad + [\mathbb{Q}\varphi_2(c,x) + \mathbb{C}_1(x)\mathbb{P}\varphi_1(c,x) + \mathbb{C}_2(x)\varphi(c)] + \mathbb{R}_2^{\varepsilon}(x)\varphi(c).
\end{aligned} \tag{[2.4.39]}$$

The first term on the right-hand side of [2.4.39] is zero because $\mathbb{Q}\varphi(c) \equiv 0$. The second term on the right-hand side of [2.4.39] is zero due to the balance condition [2.4.9]:

$$\mathbb{Q}\varphi_1(c,x) + \mathbb{C}_1(x)\varphi(c) = 0. \tag{[2.4.40]}$$

So, the solution of equation [2.4.40] is given by the following relation (see Korolyuk and Limnios 2005, Ch. 5, section 2):

$$\varphi_1(c,x) = \mathbb{R}_0\mathbb{C}_1(x)\varphi(c). \tag{[2.4.41]}$$

Next, we substitute [2.4.41] in the third term of the right-hand side of [2.4.39]. Then the solution of the singular perturbation problem

$$\mathbb{Q}\varphi_2(c,x) + \mathbb{C}_1(x)\mathbb{P}\varphi_1(c,x) + \mathbb{C}_2(x)\varphi(c) = \mathbb{L}^0\varphi(c), \tag{[2.4.42]}$$

is given by the following relation (see Korolyuk and Limnios 2005, Ch. 5, section 2, Proposition 5.2):

$$\mathbb{L}^0\varphi(c) = \Pi\,[\,\mathbb{C}_1(x)\mathbb{P}\mathbb{R}_0\mathbb{C}_1(x) + \mathbb{C}_2(x)\,]\,\varphi(c). \tag{[2.4.43]}$$

The following transformation of operator [2.4.43] specifies the form of the limit operator in Theorem 2.4.1. Given the definition of operators [2.4.29], we calculate the first term in [2.4.43]:

$$\begin{aligned}
\Pi\mathbb{C}_1(x)\mathbb{P}\mathbb{R}_0\mathbb{C}_1(x)\varphi(c) &= \Pi\mathbb{C}_1(x)\mathbb{P}\mathbb{R}_0 V(x)c\varphi'(c)\\
&= \Pi V(x)c\mathbb{P}\mathbb{R}_0 V(x)[c\varphi'(c)]'\\
&= \Pi V(x)\mathbb{P}\mathbb{R}_0 V(x)[c\varphi'(c) + c^2\varphi''(c)].
\end{aligned} \tag{[2.4.44]}$$

Let us calculate the parameter represented by the composition of the operators in the right-hand side of [2.4.44]:

$$\widehat{V}_0 := \Pi V(x)\mathbb{P}\mathbb{R}_0 V(x) = \Pi V(x)\mathbb{Q}\mathbb{R}_0 V(x) + \Pi V(x)\mathbb{R}_0 V(x). \tag{[2.4.45]}$$

Using the equality [2.4.17] as well as the balance condition [2.4.9], we obtain the relations [2.4.15] and [2.4.16].

So the right-hand side of [2.4.45] has a presentation $\widehat{V_0} = \widehat{V_0^2} - \widehat{V^2}$, that is, the first term in [2.4.43] has the form

$$[\Pi\, \mathbb{C}_1(x)\mathbb{P}\mathbb{R}_0\mathbb{C}_1(x)]\, \varphi(c) = \widehat{V_0}\, [c\varphi'(c) + c^2\varphi''(c)]. \tag{2.4.46}$$

The calculation of the second term in [2.4.43] gives the following expression:

$$\Pi\, \mathbb{C}_2(x)\, \varphi(c) = 1/2[\widehat{V^2}c^2 + \widehat{\sigma^2}]\varphi''(c). \tag{2.4.47}$$

The limit operator [2.4.43] has a representation [2.4.14]–[2.4.16], which completes the proof of Lemma 2.4.3. □

The completion of proof of Theorem 2.4.1 is realized by applying the model limit theorem (Korolyuk and Limnios 2005, Theorem 6.3), which ensures the convergence of finite-dimensional distributions [2.4.13]. □

Conclusion. The stationary nature of the limit process $\zeta(t)$, $t \geq 0$, given by the solution of the stochastic differential equation [2.4.20], provides the inequality $\widehat{V_0^2} < \widehat{V^2}$ (see [2.4.16]).

2.5. Adapted SEs

Here we consider adapted SEs with a random time change which transforms a discrete stochastic basis into a continuous basis. The adapted SEs are searched in a continuous stochastic basis in the series scheme (Koroliouk 2017). The passage to the limit, by a parameter of series, generates an approximation by diffusion processes with evolution (Jacod and Shiryaev 1987, Ch. 1).

As before, the SEs are given by averaged sums of random variables that take a finite number of possible values.

The SEs are considered in a discrete stochastic basis

$$\mathfrak{B}_{\mathbb{N}} = (\Omega,\, \mathfrak{F},\, (\mathfrak{F}_k,\, k \in \mathbb{N}),\, \mathcal{P})$$

with filtration $\mathfrak{F}_k$, $k \in \mathbb{N} = \{0, 1, ...\}$ on a probability space $(\Omega,\, \mathfrak{F},\, \mathcal{P})$.

In section 2.4, we investigated the SEs given by a random time change (Nielsen and Shiryaev 2010, Ch. 1) which transforms a discrete stochastic basis $\mathfrak{B}_{\mathbb{N}}$ into a continuous basis:

$$\mathfrak{B}_T = (\Omega, \mathfrak{G}, (\mathfrak{G}_t, t \in \mathbb{R}_+), \mathcal{P}).$$

2.5.1. *Bernoulli approximation of the SE stochastic component*

Consider SEs in a discrete stochastic basis $\mathfrak{B}_{\mathbb{N}}$. The evolutionary (predictable) component and the stochastic component are described.

The stochastic component is the martingale differences:

$$\Delta\mu_N(k+1) := \Delta S_N(k+1) - E[\Delta S_N(k+1) \,|\, S_N(k)], \quad k \geq 0. \quad [2.5.1]$$

Given the difference evolution equation [1.2.13], the martingale differences [2.5.1] have a representation

$$\Delta\mu_N(k+1) = \Delta S_N(k+1) + V_0(S_N(k)), \quad k \geq 0. \quad [2.5.2]$$

Conclusion 2.5.1. The increments of the SEs are determined by the sum of two components

$$\Delta S_N(k+1) = -V_0(S_N(k)) + \Delta\mu_N(k+1), \quad k \geq 0. \quad [2.5.3]$$

The predictable component $V_0(S_N(k))$, $k \geq 0$, is given by the RFI.

The martingale differences [2.5.1] are characterized by the first two moments (Barndorff-Nielsen and Shiryaev 2010):

$$E\Delta\mu_N(k+1) = 0,$$

$$E[(\Delta\mu_N(k+1))^2 \,|\, S_N(k)] = \sigma^2(S_N(k))/N, \quad k \geq 0. \quad [2.5.4]$$

The dispersion of the stochastic component has the form (Koroliouk 2015a):

$$\sigma^2(c) = 1 - V^2(c), \quad V(c) = c - V_0(c), \quad |c| \leq 1. \quad [2.5.5]$$

Stochastic dynamics of the SEs $S_N(k)$, $k \geq 0$, is given by the SDEs [2.5.3]–[2.5.5].

Properties of the stochastic component are allowed specification.

LEMMA 2.5.1.– *The stochastic component determined by the martingale differences* [2.5.2] *have the following representation:*

$$\Delta\mu_N(k+1) = \frac{1}{N}\sum_{n=1}^{N}\beta_n(k+1), \quad k \geq 0. \tag{2.5.6}$$

The sample variables $\beta_n(k+1)$, $0 \leq n \leq N$, $k \geq 0$, *take two values:*

$$\beta_n(k+1) = \Big\{\pm 1 - V(C(k)) \text{ with probability } P_\pm(k+1)\Big\}, \tag{2.5.7}$$

where

$$P_\pm(k+1) = 1/2\big[1 \pm C(k+1)\big] = \frac{1}{2}\big[1 \pm V(C(k))\big],$$
$$C(k) := P_+(k) - P_-(k), \quad k \geq 0.$$

Proof of Lemma 2.5.1.– Since

$$\Delta S_N(k+1) = S_N(k+1) - S_N(k) = \frac{1}{N}\sum_{n=1}^{N}[\delta_n(k+1) - \delta_n(k)],$$

then

$$\Delta\mu_N(k+1) = \frac{1}{N}\sum_{n=1}^{N}\beta_n(k+1) = \frac{1}{N}\sum_{n=1}^{N}[\delta_n(k+1) - \delta_n(k)] + V_0(\Delta S_N(k)).$$

Taking into account that

$$\beta_n(k+1) = \delta_n(k+1) - \delta_n(k) + V_0(\delta_n^0(k)) = \delta_n(k+1) - \delta_n^0(k),$$
$$\delta_n^0(k) := \delta_n(k) - V_0(\delta_n^0(k)), \quad E[\delta_n^0(k) \mid S_N(k) = c] = V(c),$$

and also, for all n:

$$E[\delta_n(k+1) \mid S_N(k) = c] = P_+(k+1) - P_-(k+1) = C(k+1) = V(C(k)),$$

we have the values of the first two moments $\forall c$: $|c| \leq 1$,

$$E[\beta_n(k+1) \mid S_N(k) = c] = E[\delta_n(k+1) \mid S_N(k) = c] - V(c) \equiv 0,$$
$$E[\beta_n^2(k+1) \mid S_N(k) = c]$$
$$= [1 - V(c)]^2\, P_+(k+1) + [1 + V(c)]^2\, P_-(k+1) = 1 - V^2(c),$$

which completes the proof of Lemma 2.5.1. □

Conclusion 2.5.2. The stochastic component [2.5.6] is given by the Bernoulli distribution:

$$\begin{aligned} B_N(\nu; V(C(k))) &= P\{\Delta\mu_N(k+1) = \nu - V(C(k)) \mid S_N(k) = C(k)\} \\ &= \frac{N!}{N_+!N_-!} P_+^{N_+}(k+1) P_-^{N_-}(k+1), \end{aligned} \qquad [2.5.8]$$

where

$$N_\pm/N = 1/2[1 \pm \nu], \quad \nu = \nu_+ - \nu_-, \quad \nu_\pm = N_\pm/N.$$

with the first two moments:

$$E[\beta_n(k+1) \mid S_N(k)] = 0 \quad \forall k \geq 0,$$
$$E[\beta_n^2(k+1) \mid S_N(k)] = \sigma^2(S_N(k)) = 1 - V^2(S_N(k)).$$

Now the SE dynamics has two interpretations:

– The increments of SEs $\Delta S_N(k)$ are determined by the difference equation [2.5.3], in which the stochastic component is given by the Bernoulli distribution [2.5.8];

– The probabilities [2.5.7] determine the Bernoulli distribution [2.5.8] of the stochastic component, at a given kth stage.

2.5.2. *Adapted SEs*

The passage from discrete stochastic basis $\mathfrak{B}_\mathbb{N} = (\Omega, \mathfrak{F}, (\mathfrak{F}_k, k \in \mathbb{N}), \mathcal{P})$ to continuous stochastic basis $\mathfrak{B}_T = (\Omega, \mathfrak{G}, (\mathfrak{G}_t, t \in \mathbb{R}_+), \mathcal{P})$ is realized by a *random time change*

$$\nu(t), t \geq 0, \quad \nu(0) = 0.$$

The increasing process $\nu(t)$, $t \geq 0$, right continuous and having left limits, is given by the Markov moments of jumps:

$$\tau_k := \inf\{t : \nu(t) \geq k\}, \quad k \in \mathbb{N}.$$

Regularity of the process $\nu(t)$, $t \geq 0$, is ensured by the condition

$$P\{\tau_k < +\infty\} \quad \forall k > 0. \qquad [2.5.9]$$

The counting renewal process $\nu(t)$, $t \geq 0$, is considered (for simplicity) as a stationary Poisson process with exponentially distributed renewal intervals $\theta_{k+1} := \tau_{k+1} - \tau_k$, $k \geq 0$:

$$P\{\theta_{k+1} \geq t\} = \exp(-qt), \quad 0 < q < +\infty, \quad t \geq 0.$$

It is known (Korolyuk 1999) that the compensator of the Poisson process $\nu(t)$, $t \geq 0$, is given by

$$E\nu(t) = qt, \quad t \geq 0.$$

DEFINITION 2.5.1.– The random time change in a discrete basis $\mathfrak{B}_{\mathbb{N}}$ is given by the filtration

$$\mathfrak{G}_t = \mathfrak{F}_{\nu(t)}, \quad t \geq 0. \qquad [2.5.10]$$

According to Lemma 3.8 (Jacod and Shiryaev 1987, Ch. 2, section 3b), there are the properties: $\tau_0 = 0$, $\mathfrak{G}_0 = \mathfrak{F}_0$, and $\mathfrak{G}_{\tau_k} = \mathfrak{F}_k$ on the set $\tau_k < \infty$ and $\mathfrak{F}_{k-1} = \mathfrak{G}_{\tau_k -}$, $k > 0$.

In particular, if $\nu(t) = [t]$ be integer part of a positive number $t > 0$, then the basis $\mathfrak{B}_T$ coincides with the basis $\mathfrak{B}_{\mathbb{N}}$.

DEFINITION 2.5.2.– The adapted SE with a random time change [2.5.10] is given by

$$\alpha_N(t) := S_N(\nu(t)), \quad t \geq 0,$$

or

$$\alpha_N(t) := S_N(k), \quad \tau_k \leq t < \tau_{k+1}, \quad t \geq 0.$$

Conclusion 2.5.3. The adapted SE is a *special semimartingale*, for which:

– the predictable component, given by the RFI $V_0(c)$, $|c| \leq 1$;

– the stochastic component is given by the Bernoulli distribution [2.5.8] of increments $\Delta\mu_N(k+1)$, $k \geq 0$.

We obtain the following representation (see equation [2.5.3]):

$$\alpha_N(t) = \alpha_N(0) + V_N(t) + M_N(t), \quad t \geq 0,$$

$$V_N(t) := -\sum_{k=0}^{\nu(t)} V_0(\alpha_N(\tau_k)), \quad M_N(t) := \sum_{k=0}^{\nu(t)} \Delta\mu_N(k+1).$$

2.5.3. *Adapted SEs in a series scheme*

The adapted SEs in a series scheme with a small parameter $\varepsilon \to 0$ ($\varepsilon > 0$) are determined by the random time change

$$\nu_\varepsilon(t) := \nu\left(\frac{t}{\varepsilon^2}\right), \quad t \geq 0, \qquad [2.5.11]$$

as well as by the following normalization of increments [2.5.3] by a series parameter ε^2.

DEFINITION 2.5.3.– The adapted SEs with a random time change [2.5.11] in series scheme

$$\alpha^\varepsilon(t) := S_N(\nu_\varepsilon(t)), \quad \varepsilon^2 = \frac{1}{N}, \quad t \geq 0,$$

are determined by the following three predictable characteristics (Jacod and Shiryaev 1987, Ch. 2):

– evolutionary component:

$$V_t^\varepsilon := -\varepsilon^2 \sum_{k=0}^{\nu_\varepsilon(t)} V_0(\alpha^\varepsilon(\tau_k^\varepsilon)), \quad t \geq 0; \qquad [2.5.12]$$

– dispersion of stochastic component:

$$\sigma_t^\varepsilon := \varepsilon^2 \sum_{k=0}^{\nu_\varepsilon(t)} \sigma^2(\alpha^\varepsilon(\tau_k^\varepsilon)), \quad \sigma^2(c) := 1 - V^2(c), \quad t \geq 0, \qquad [2.5.13]$$

– compensating measure of jumps:

$$\Gamma_t^\varepsilon(g) := \varepsilon^2 \sum_{k=0}^{\nu_\varepsilon(t)} E\big[g(\alpha^\varepsilon(\tau_{k+1}^\varepsilon)) \,|\, \mathfrak{G}_{\tau_k^\varepsilon}^\varepsilon\big], \quad g \in C_2(\mathbb{R}), \quad t \geq 0. \qquad [2.5.14]$$

Here the class of functions $C_2(R)$ is defined in Jacod and Shiryaev (1987); the filtration in the condition [2.5.14] is $\mathfrak{G}_t^\varepsilon := \mathfrak{F}_{\nu_\varepsilon(t)}^\varepsilon$, $t \geq 0$ (see equation [2.5.10]).

The predictable characteristics of the adapted SE [2.5.12]–[2.5.14] depend on the current value of the SEs $\alpha^\varepsilon(\tau_k^\varepsilon)$, $k \geq 0$, at renewal moments $\tau_k^\varepsilon = \varepsilon^2 \tau_k$, $k \geq 0$, of the random time change $\nu_\varepsilon(t)$, $t \geq 0$. So the study of the convergence problem for adapted SEs in the series scheme, by $\varepsilon \to 0$, is realized in two stages (see Liptser 1994).

Stage 1. Determination of the compactness conditions of the adapted SEs $\alpha^\varepsilon(t), 0 \le t \le T, \varepsilon \ge 0$.

Stage 2. Under additional conditions on the predictable characteristics, that is, functions $V_0(c)$, $\sigma^2(c)$, $|c| \le 1$, identify the limit process given by the corresponding predictable characteristics.

At first stage, the approach proposed in the work Liptser (1994) is used (see also Limnios and Samoilenko 2013). Namely, the condition of *compact embedding* takes place:

$$\lim_{c\to\infty} \sup_{\varepsilon>0} P\{ \sup_{0\le t\le T} |\alpha^\varepsilon(t)| > c\} = 0. \qquad [2.5.15]$$

LEMMA 2.5.2.– *By the condition on the initial values* $E|\alpha^\varepsilon(0)| \le c_0$ *with a constant that does not depend on* ε, *the following estimate takes place:*

$$E \sup_{0\le t\le T} |\alpha^\varepsilon(t)|^2 \le C. \qquad [2.5.16]$$

PROOF 2.26.– The semimartingale representation of the adapted SEs and the properties of the predictable characteristics [2.5.12]–[2.5.14] are used.

The evolutionary component V_t^ε, $t \ge 0$, has an estimate:

$$\sup_{0\le t\le T} |V_t^\varepsilon| \le C_V \int_0^T [1 + |V_0(\alpha_*^\varepsilon(t))|] \, dt,$$

$$\alpha_*^\varepsilon(t) := \sup_{0\le s\le t} |\alpha^\varepsilon(s)|.$$

The dispersion of the stochastic component M_t^ε, $t \ge 0$, is uniformly bounded in mean-square:

$$E \sup_{0\le t\le T} |M_t^\varepsilon|^2 \le C_\sigma \int_0^T [1 + \sigma^2(\alpha_*^\varepsilon(t))] \, dt.$$

So, for the adapted SEs, we obtain the following estimate:

$$E \sup_{0\le t\le T} |\alpha^\varepsilon(t)|^2 \le 3E\left[[\alpha^\varepsilon(0)]^2 + \sup_{0\le t\le T} |V_t^\varepsilon| + \sup_{0\le t\le T} |M_t^\varepsilon|^2 \right].$$

The last three inequalities provide the following estimate:

$$E[\alpha_*^\varepsilon(T)]^2 \leq C_1 + C_2 \int_0^T E|\alpha_*^\varepsilon(t)|^2 \, dt.$$

Finally, the last inequality—Gronwall-Bellman—establishes the inequality [2.5.16]. Lemma 2.5.2 is proved. □

Now Kolmogorov's inequality for adapted SEs $\alpha^\varepsilon(t)$, $0 \leq t \leq T$, $\varepsilon \geq 0$, establishes the condition of compact embedding [2.5.15].

REMARK 2.5.1.– Another approach to establish the condition of compact embedding [2.5.15] is shown in the monograph (Ethier and Kurtz 1986, Ch. 4, section 5).

Conclusion 2.5.4. Under Lemma 2.5.2, we have the following estimate:

$$E|\alpha^\varepsilon(t) - \alpha^\varepsilon(t')|^2 \leq C_T|t - t'|, \quad 0 \leq t, t' \leq T. \qquad [2.5.17]$$

In conditions [2.5.15] and [2.5.17], there is a compactness of processes $\alpha^\varepsilon(t), 0 \leq t \leq T, \varepsilon > 0$.

In the second stage, under the condition of compactness of the adapted SEs in the series scheme $\alpha^\varepsilon(t)$, $0 \leq t \leq T$, $\varepsilon > 0$, the verification of limit process is realized by the study of convergence (by $\varepsilon \to 0$) of predictable characteristics [2.5.12]–[2.5.14]. First of all, the convergence of the compensating measure of jumps [2.5.14] is established:

$$\sup_{0 \leq t \leq T} \Gamma_t^\varepsilon(g) \xrightarrow{D} 0, \quad \varepsilon \to 0,$$

which is provided by the Lindeberg condition:

$$\sum_{n=1}^{N} E[\beta_n^\varepsilon(k+1)]^2 \cdot I(|\beta_n(k+1)|) \geq \frac{h}{\varepsilon} \, S_N(k) = c] \to 0, \quad \varepsilon \to 0,$$

for the sample sum

$$\Delta\mu^\varepsilon(k+1) := \varepsilon \sum_{n=1}^{N} \beta_n(k+1),$$

with dispersion

$$D_N^2 := E[(\Delta\mu^\varepsilon(k+1))^2 \mid S_N(k) = c] = \sigma^2(c).$$

Next, we need to prove the convergence of the evolutionary component and the dispersion of the stochastic component of the semimartingale $\alpha^\varepsilon(t), t \geq 0$.

LEMMA 2.5.3.– *In the conditions of Lemma 2.5.2, the convergences in distribution takes place, as* $\varepsilon \to 0$*:*

$$V_t^\varepsilon \xrightarrow{D} V_t^0 = -\int_0^{qt} V_0(\alpha^0(u))\,du, \quad 0 \leq t \leq T,$$

$$\sigma_t^\varepsilon \xrightarrow{D} \sigma_t^0 = \int_0^{qt} \sigma^2(\alpha^0(u))\,du, \quad 0 \leq t \leq T, \quad \sigma^2(c) = 1 - V^2(c).$$

Here, the limit process $\alpha^0(t)$, $t \geq 0$, is determined by the condition of compactness (see Lemma 2.5.2):

$$\alpha^{\varepsilon_r}(t) \xrightarrow{D} \alpha^0(t), \quad \varepsilon_r \to 0, \quad r \to \infty.$$

Proof of Lemma 2.5.3.– Since both predictable characteristics [2.5.12] and [2.5.13] have the same structure of the integral functional on the process $\alpha^\varepsilon(t), t \geq 0$, it is sufficient to investigate the convergence of one of them, for example, the evolutionary component [2.5.12].

We use the martingale characterization

$$\mu_t^\varepsilon = \varphi(V_t^\varepsilon) - \int_0^t L_V^\varepsilon \varphi(V_u^\varepsilon)\,du, \quad t \geq 0.$$

The integral functional generator of [2.5.12] is

$$L_V^\varepsilon \varphi(c) = \varepsilon^{-2} q[\varphi(c - \varepsilon^2 V_0(c)) - \varphi(c)], \quad \varphi(c) \in C^2(\mathbb{R}).$$

It admits an asymptotic representation on a class of test functions:

$$L_V^\varepsilon \varphi(c) = L_V^0 \varphi(c) + R_\varepsilon \varphi(c), \quad \varphi(c) \in C^2(\mathbb{R})$$

with a neglectable term

$$R_\varepsilon \varphi(c) \to 0, \quad \varepsilon \to 0, \quad \varphi(c) \in C^2(\mathbb{R}).$$

The limit operator L_V^0 determines the evolution

$$L_V^0\varphi(c) = -qV_0(c)\varphi'(c), \quad \varphi(c) \in C^2(\mathbb{R}).$$

The limit evolution is given by

$$V_t^0 = -\int_0^{qt} V_0(\alpha^0(u))\, du, \quad t \geq 0.$$

Consequently, the compactness of the adapted SE [2.5.16], established by Lemma 2.5.2, ensures the convergence of martingales:

$$\mu_t^\varepsilon \Rightarrow \mu_t^0 = \varphi(V_t^0) - \int_0^t L_V^0\varphi(V_u^0)\, du, \quad 0 \leq t \leq T, \quad \varepsilon \to 0.$$

Similarly, the convergence of quadratic characteristic [2.5.13] is established using the martingale characterization:

$$\mu_t^\varepsilon = \varphi(\sigma_t^\varepsilon) - \int_0^t L_\sigma^\varepsilon\varphi(\sigma_u^\varepsilon)\, du, \quad t \geq 0,$$

with the generator

$$L_\sigma^\varepsilon\varphi(c) = \varepsilon^{-2}q[\varphi(c + \varepsilon^2\sigma_0^2(c)) - \varphi(c)], \quad \varphi(c) \in C^2(\mathbb{R}),$$

which admits an asymptotic representation on the class of test functions $\varphi(c) \in C^2(\mathbb{R})$:

$$L_\sigma^\varepsilon\varphi(c) = q\sigma_0^2(c)\varphi'(c) + R_\varepsilon\varphi(c), \quad \varphi(c) \in C^2(\mathbb{R})$$

with a neglectable term

$$R_\varepsilon\varphi(c) \to 0, \quad \varepsilon \to 0, \quad \varphi(c) \in C^2(\mathbb{R}).$$

Consequently, the limit quadratic characteristic has a representation:

$$\sigma_t^0 = \int_0^{qt} \sigma^2(\alpha_0(u))\, du, \quad t \geq 0, \quad \sigma^2(c) = 1 - V^2(c).$$

In the final stage, the condition of uniqueness of semimartingale characterization of the diffusion MP with evolution $\alpha^0(t)$, $t \geq 0$, is used, given by the generator (Jacod and Shiryaev 1987, Ch. 9):

$$L_V^0\varphi(c) = -V_0(c)\varphi'(c) + 1/2\sigma^2(c)\varphi''(c), \quad \varphi(c) \in C^2(\mathbb{R}).$$

□

THEOREM 2.5.1.– *The adapted SE $\alpha^\varepsilon(t)$, $t \geq 0$, in the series scheme with the small parameter $\varepsilon \to 0$ ($\varepsilon > 0$), given by predictable characteristics* [2.5.12]–[2.5.14], *under the additional condition of convergence of the initial values:*

$$\alpha^\varepsilon(0) \xrightarrow{D} \alpha_0, \quad E\alpha^\varepsilon(0) \to E\alpha_0, \quad \varepsilon \to 0.$$

converges, in distribution, to a diffusion process with evolution with a random time change

$$\alpha^\varepsilon(t) \xrightarrow{D} \alpha^0(t), \quad 0 \leq t \leq T, \quad \varepsilon \to 0.$$

In this case, the predictable characteristics of limit process $\alpha^0(t)$, $t \geq 0$, have the representation:

$$V_t^0 = \int_0^{qt} V_0(\alpha^0(u))\, du, \quad \sigma_t^0 = \int_0^{qt} \sigma^2(\alpha^0(u))\, du, \quad 0 \leq t \leq T,$$

and there is no compensating jump measure:

$$B_t^\varepsilon(g) \to 0, \quad \varepsilon \to 0, \quad g(c) \in C_1(\mathbb{R}).$$

Conclusion 2.5.5. The limit diffusion process with evolution $\alpha^0(t)$, $t \geq 0$, is given by the following stochastic differential equation:

$$d\alpha(t) = -V_0(\alpha(t)), dt + \sigma(\alpha(t))\, dW_t, \quad t \geq 0,$$

with the scaling time change

$$\alpha^0(t) = \alpha(qt), \quad t \geq 0.$$

The intensity q is determined by the equality $E\nu(t) = qt$.

2.6. DMPs in an asymptotical diffusion environment

The asymptotic diffusion environment given by the solution of the DEE in an ergodic Markov environment provided the limit diffusion environment for the limit DMP.

2.6.1. *Asymptotic diffusion perturbation*

The MRE is given by a two-component MC (x_n, τ_n), $n \geq 0$, homogeneous by the second component (Korolyuk and Limnios 2005):

$$\mathcal{P}\{x_{n+1} \in B, \, \tau_{n+1} \leq t + c \,|\, x_n = x, \, \tau_n = c\} = P(x, B)G_x(t), \; t \geq 0. \qquad [2.6.1]$$

The semi-Markov kernel

$$Q(x, B, t) := P(x, B)G_x(t), \quad x \in E, \quad B \in \mathcal{E}, \quad t \geq 0, \qquad [2.6.2]$$

specifies the transition probabilities of the two-component MC (x_n, τ_n), $n \geq 0$:

$$P(x, B) := \mathcal{P}\{x_{n+1} \in B \,|\, x_n = x\}, \quad x \in E, \quad B \in \mathcal{E}, \qquad [2.6.3]$$

$$G_x(t) := \mathcal{P}\{\theta_x \leq t\} = \mathcal{P}\{\theta_{n+1} \leq t \,|\, x_n = x\}, \quad x \in E, \quad B \in \mathcal{E}, \quad t \geq 0. \qquad [2.6.4]$$

Here, the random variables

$$\theta_{n+1} := \tau_{n+1} - \tau_n, \quad n \geq 0, \quad \tau_0 = 0, \qquad [2.6.5]$$

determine the sojourn time in the states $x \in E$ of the embedded MC x_n, $n \geq 0$. In the particular case of exponentiality of distributions [2.6.4],

$$\overline{G}_x(t) = 1 - G_x(t) = e^{-q(x)t}, \quad t \geq 0.$$

The MRE is really Markov, that is, x_{τ_n}, $n \geq 0$, is an MC.

For arbitrary functions of distribution of sojourn time [2.6.4], the MRE becomes a *semi-Markov* process.

The asymptotic diffusion is generated by the following the DEE:

$$\Delta Y^\varepsilon(t + \varepsilon^2) = \varepsilon A_0(Y^\varepsilon(t); \, x) + \varepsilon^2 A(Y^\varepsilon(t); \, x), \quad x \in E. \qquad [2.6.6]$$

The singular term $A_0(y; \, x)$ with the scaling factor ε satisfies the balance condition

$$\int_E \rho(dx) A_0(y; \, x) \equiv 0. \qquad [2.6.7]$$

The averaging [2.6.7] is determined by the stationary distribution of the embedded MC x_n, $n \geq 0$:

$$\rho(B) = \int_E \rho(\mathrm{d}x) P(x, B), \quad B \in \mathcal{E}, \quad \rho(E) = 1. \qquad [2.6.8]$$

It is known (Korolyuk and Limnios 2005), that the solution of DEEs [2.6.6] and [2.6.7] generates a diffusion process, as $\varepsilon \to 0$.

According to the tradition of singular perturbation theory (Korolyuk and Limnios 2005), DMP is considered in a discrete–continuous time with a small parameter $\varepsilon \to 0$ ($\varepsilon > 0$): $t_k = [k\varepsilon^2]$, $k \geq 0$.

Basic assumption 1. The DMP with asymptotic diffusion is given by the SDE solution

$$\Delta\zeta^\varepsilon(t + \varepsilon^2) = -\varepsilon^2 V(Y^\varepsilon(t))\zeta^\varepsilon(t) + \varepsilon\sigma(Y^\varepsilon(t))\Delta\mu^\varepsilon(t), \quad t \geq 0. \quad [2.6.9]$$

The asymptotic diffusion $Y^\varepsilon(t)$, $t \geq 0$, is determined by the DEE [2.6.6] solution with the balance condition [2.6.7].

The DMP $\zeta^\varepsilon(t)$, $t \geq 0$, given by the SDE solution [2.6.9], is supplemented by an asymptotic diffusion $Y^\varepsilon(t)$, $t \geq 0$, given by the DEE solution [2.6.6].

THEOREM 2.6.1.– *In the uniform ergodicity conditions of the embedded MC x_n, $n \geq 0$, with the stationary distribution* [2.6.8], *the two-component process of the DMP* [2.6.9], *and the asymptotic diffusion $Y^\varepsilon(t)$, $t \geq 0$, converge in distribution, as $\varepsilon \to 0$, to an Ornstein–Uhlenbeck diffusion process with evolution:*

$$\{\zeta^\varepsilon(t), Y^\varepsilon(t)\} \xrightarrow{D} \{\zeta^0(t), Y^0(t)\}, \quad \varepsilon \to 0, \quad 0 \leq t \leq T. \qquad [2.6.10]$$

The limit two-component diffusion process $\{\zeta^0(t), Y^0(t)\}$, $t \geq 0$, with evolution, is determined by the generator

$$\begin{aligned} \mathfrak{L}^0(y)\varphi(c, y) = &- V(y)\varphi'_c(c, y) + \frac{1}{2}\sigma^2(y)\varphi''_c(c, y) \\ &+ \widehat{A}(y)\varphi'_y(c, y) + \frac{1}{2}\widehat{B}^2(y)\varphi''_y(c, y), \ \varphi \in C_0^2(R, \mathcal{B}). \end{aligned} \qquad [2.6.11]$$

Here by definition

$$\widehat{A}(y) := \int_E \rho(\mathrm{d}x)[A(y;\, x) + A_1(y;\, x)], \tag{2.6.12}$$

$$A_1(y;\, x) := A_0(y;\, x)\, P\, R_0\, A'_{0y}(y;\, x), \tag{2.6.13}$$

$$\widehat{B}^2(y) = \int_E \rho(\mathrm{d}x) B(y;\, x),$$

$$B(y;\, x) = A_0(y;\, x)\, P\, [R_0 + 1/2\mathbb{I}]\, A_0(y;\, x). \tag{2.6.14}$$

The potential kernel (Korolyuk and Limnios 2005, Section 5.2):

$$R_0 = (Q + \Pi)^{-1} - \Pi, \quad Q := P - I, \quad \Pi\varphi(x) := \int_E \rho(\mathrm{d}x)\varphi(x). \tag{2.6.15}$$

REMARK 2.6.1.– With an additional condition for the continuity of parameters $V(y)$ and $G(y)$, $y \in R$, the MPs $\zeta^0(t)$, $Y^0(t)$, $t \geq 0$, given by the generator [2.6.11], are Feller MPs (Liggett 2010, p. 93, Theorem 3.3).

REMARK 2.6.2.– The limit two-component diffusion process $(\zeta^0(t),\, Y^0(t),\, t \geq 0)$, determined by the generator [2.6.11], has a stochastic presentation by SDEs:

$$\begin{aligned} \mathrm{d}\zeta^0(t) &= -V(Y^0(t))\zeta^0(t)\, \mathrm{d}t + \sigma(Y^0(t))\, \mathrm{d}W(t), \\ \mathrm{d}Y^0(t) &= \widehat{A}(Y^0(t))\, \mathrm{d}t + \widehat{B}(Y^0(t))\, \mathrm{d}W_0(t). \end{aligned} \tag{2.6.16}$$

The parameters of limit diffusion $\zeta^0(t)$, $t \geq 0$, depend on the diffusion process $Y^0(t)$, $t \geq 0$, which is unequally determined by the averaged values $\widehat{A}$, $\widehat{B}$ [2.6.12]–[2.6.15] that satisfy the uniform Lipshitz conditions (Gihman and Skorohod 1979, Ch. 2).

Proof of Theorem 2.6.1.– The basic idea of the proof is formulated in the beginning of section 2.1.2. The operator characterization of MP, determined by the generator on the class of real-valued functions, acting on the set of values of the MP, is used (Korolyuk and Limnios 2005).

First of all, we used the operator characterization of the extended three-component MC

$$\{\zeta^\varepsilon(t_n),\, Y^\varepsilon(t_n),\, x^\varepsilon(t_n) := x_{t_n/\varepsilon^2}\}, \quad t_n = n\varepsilon^2, \quad \varepsilon > 0, \tag{2.6.17}$$

in the following form.

LEMMA 2.6.1.– *The extended MC is determined by the generator*

$$\mathbb{L}^{\varepsilon}(x)\varphi(c,\, y,\, x) = \varepsilon^{-2}[\Gamma^{\varepsilon}(y)A^{\varepsilon}(x)P - \mathbb{I}]\varphi(c,\, y,\, x). \qquad [2.6.18]$$

Here the following operators are used:

$$\begin{aligned} \Gamma^{\varepsilon}(y)\varphi(c) &:= E[\varphi(c + \Delta\zeta^{\varepsilon}(t;\, y)) \,|\, \zeta^{\varepsilon}(t;\, y) = c], \\ A^{\varepsilon}(x)\varphi(y) &:= E[\varphi(y + \Delta Y^{\varepsilon}(t;\, x)) \,|\, Y^{\varepsilon}(t;\, x) = y], \qquad [2.6.19] \\ P\varphi(x) &:= \int_E P(x,\, dz)\varphi(z). \end{aligned}$$

The statement of Lemma 2.6.1 follows from the definition of the generator of three-component MC [2.6.17].

A significant step of the proof of Theorem 2.6.1 is realized in the next lemma.

LEMMA 2.6.2.– *The generator* [2.6.18] *and* [2.6.19] *of a three-component MC* [2.6.17], *acting on real-valued functions* $\varphi(c,\, y,\, x)$, *which have finite derivatives up to the third order, admits the following asymptotic representation:*

$$\begin{aligned} &\mathbb{L}^{\varepsilon}(x)\varphi(c,\, y,\, x) \\ &= \left[\varepsilon^{-2}\mathbb{Q} + \varepsilon^{-1}\mathbb{A}_0(x)P + \mathbb{A}(x)P + \mathbb{L}^0(y)P + \mathbb{R}_{\varepsilon}(y;\, x)\right]\varphi(c,\, y,\, x), \end{aligned} \qquad [2.6.20]$$

$$\mathbb{A}_0(x)\varphi(y) = A_0(y;\, x)\varphi'(y), \quad \mathbb{A}(x)\varphi(y) = A(y;\, x)\varphi'(y), \qquad [2.6.21]$$

$$\mathbb{L}^0(y)\varphi(c) = -V(y)c\varphi'(c) + \frac{1}{2}\sigma^2(y)\varphi''(c). \qquad [2.6.22]$$

The residual term $\mathbb{R}_{\varepsilon}(y;\, x)$ *is neglectable:*

$$\mathbb{R}_{\varepsilon}(y;\, x)\varphi(c,\, y,\, x) \to 0, \quad \varepsilon \to 0, \quad \varphi \in C^3(R^2).$$

Proof of Lemma 2.6.2.– We use the following transformation of the generator [2.6.18]:

$$\begin{aligned} &\mathbb{L}^{\varepsilon}(x)\varphi(c,\, y,\, x) \\ &= \varepsilon^{-2}\left[\mathbb{Q} + (A^{\varepsilon} - \mathbb{I})P + (\Gamma^{\varepsilon} - \mathbb{I})P + \mathbb{R}_{\varepsilon}(y;\, x)\right]\varphi(c,\, y,\, x). \end{aligned} \qquad [2.6.23]$$

The residual term is

$$\mathbb{R}_\varepsilon(y;\, x)\varphi(c,\, y,\, x) = (\Gamma^\varepsilon - \mathbb{I})(A^\varepsilon - \mathbb{I})P\varphi(c,\, y,\, x). \qquad [2.6.24]$$

Then we calculate

$$\begin{aligned}\varepsilon^{-2}[\Gamma^\varepsilon(y) - \mathbb{I}]\varphi(c) &= \varepsilon^{-2}E[\varphi(c + \Delta\zeta^\varepsilon(t;\, y) \mid \zeta^\varepsilon(t;\, y) = c] \\ &= [\mathbb{L}^0(y) + \mathbb{R}_\varepsilon(y;\, c)]\varphi(c).\end{aligned} \qquad [2.6.25]$$

The limit generator $\mathbb{L}^0(y)$ is defined in [2.6.22].

The next transformation

$$\begin{aligned}&\varepsilon^{-2}[A^\varepsilon(x) - \mathbb{I}]\varphi(y) \\ &= \varepsilon^{-2}E[\varphi(y + \Delta y^\varepsilon(t;\, x)) - \varphi(y) \mid y^\varepsilon(t;\, x) = y] \\ &= \varepsilon^{-2}\left[E[\Delta y^\varepsilon(t;\, x)]\varphi'(y) + \frac{1}{2}E[\Delta y^\varepsilon(t;\, x)]^2\varphi''(y) + \mathbb{R}_\varepsilon(y;\, c)\right] \\ &= \left[\varepsilon^{-1}\mathbb{A}_0(y;\, x) + \mathbb{A}(y;\, x)\right]\varphi'(y) + \frac{1}{2}\mathbb{A}_0^2(y;\, x)\varphi''(y) + \mathbb{R}_\varepsilon\varphi(y).\end{aligned} \qquad [2.6.26]$$

enables us to get the asymptotic expansion for Lemma 2.6.2:

$$\begin{aligned}\mathbb{L}^\varepsilon(x)\varphi(c,\, y,\, x) =& [\varepsilon^{-2}\mathbb{Q} + \varepsilon^{-1}\mathbb{A}_0(x)\,P + \mathbb{A}(x)\,P + \frac{1}{2}(\mathbb{A}_0(x)\,P)^2 \\ &+ \mathbb{L}^0(y)]\varphi(c,\, y,\, x) + \mathbb{R}_\varepsilon(y;\, x)\varphi(c,\, y,\, x).\end{aligned} \qquad [2.6.27]$$

□

Next we use the solution of singular perturbation problem for the truncated operator.

LEMMA 2.6.3.– *The solution of the singular perturbation problem for the truncated operator is realized on the perturbed test functions as follows:*

$$\begin{aligned}\mathbb{L}_0^\varepsilon(x)\varphi^\varepsilon(c,\, y,\, x) &= \left[\varepsilon^{-2}\mathbb{Q} + \varepsilon^{-1}\mathbb{A}_0(x)\,P + \mathbb{A}(x)\,P + \frac{1}{2}(\mathbb{A}_0(x)\,P)^2 + \mathbb{L}^0(y)\right] \\ &\quad \times [\varphi(c,\, y) + \varepsilon\varphi_1(c,\, y,\, x) + \varepsilon^2\varphi_2(c,\, y,\, x)] \\ &= \mathfrak{L}^0(y)\varphi(c,\, y) + \mathbb{R}_\varepsilon(x)\varphi(c,\, y).\end{aligned} \qquad [2.6.28]$$

The corresponding averaged parameters are determined by the formulae [2.6.12]–[2.6.15].

The limit operator is calculated as

$$\mathfrak{L}^0(y)\varphi(c,\, y) = \mathbb{L}^0(y) + \widehat{A}(y)\varphi'_y(c,\, y) + 1/2\widehat{B}^2(y)\varphi''_y(c,\, y). \quad [2.6.29]$$

Proof of Lemma 2.6.3.– For solving the singular perturbation problem for the truncated operator, consider the following asymptotic expansion in powers of ε:

$$\begin{aligned}\mathbb{L}_0^\varepsilon(x)\varphi^\varepsilon(c,\, y,\, x) &= \varepsilon^{-2}\mathbb{Q}\varphi(c,\, y) + \varepsilon^{-1}[\mathbb{Q}\varphi_1(c,\, y,\, x) + \mathbb{A}_0(x)\varphi(c,\, y)] \\ &\quad + \Big[\mathbb{Q}\varphi_2(c,\, y,\, x) + \mathbb{A}_0(x)\, P\,\varphi_1(c,\, y,\, x) \\ &\quad + [\mathbb{A}(x) + 1/2(\mathbb{A}_0(x)P)^2 + \mathbb{L}^0(y)]\varphi(c,\, y)\Big] + \mathbb{R}_\varepsilon(x)\varphi(c,\, y).\end{aligned} \quad [2.6.30]$$

Obviously, $\mathbb{Q}\varphi(c,\, y) = 0$.

The balance condition [2.6.7] is then used. The solution of the equation

$$\mathbb{Q}\varphi_1(c,\, y,\, x) + \mathbb{A}_0(x)\varphi(c,\, y) = 0$$

is given by (Korolyuk and Limnios 2005, Section 5.4):

$$\varphi_1(c,\, y,\, x) = \mathbb{R}_0\mathbb{A}_0(x)\varphi(c,\, y).$$

Substituting this result in the following equation:

$$\mathbb{Q}\varphi_2(c,\, y,\, x) + \big[\mathbb{B}(x) + \mathbb{A}(x) + \mathbb{L}^0(y)\big]\varphi(c,\, y) = \mathfrak{L}^0(y)\varphi(c,\, y). \quad [2.6.31]$$

Here, by definition,

$$\mathbb{B}(x) := \mathbb{A}_0(x)PR_0\mathbb{A}_0(x) + 1/2(\mathbb{A}_0(x)P)^2. \quad [2.6.32]$$

The limit operator is calculated by using the balance condition

$$\mathfrak{L}^0(y)\varphi(c,\, y) = \Big\{\Pi\Big[\mathbb{B}(x) + \mathbb{A}(x)\Big]\Pi + \mathbb{L}^0(y)\Big\}\varphi(c,\, y). \quad [2.6.33]$$

Recall the projector's operation:

$$\Pi\mathbb{B}(x) = \int_E \rho(\mathrm{d}x)B(y,\, x),$$

$$B(y,\, x) = A_0(y,\, x)PR_0A_0(y,\, x) + 1/2(A_0(y,\, x))^2.$$

Taking into account the definition of the evolution operators [2.6.21], the limit generator is determined using formula [2.6.30]. □

The proof of Theorem 2.6.1 is complete. □

REMARK 2.6.3.– Using the definition of the potential operator R_0 (see Korolyuk and Limnios 2005, Section 5.5), we offer the reader a further analysis of the formula for the diffusion parameter $\mathbb{B}(x)$.

2.7. A DMP with ASD

The DMP in discrete–continuous time is given by the solution of the SDEs in the series scheme, approximated by an Ornstein–Uhlenbeck process with continuous time and ASD (see section 2.7.1).

The EG for DMP is derived from the method of large deviations analysis for MPs similarly as in the Feng–Kurtz monograph (Feng and Kurtz 2006) (see section 2.7.2).

The AF for DMP is introduced using the method developed in Freidlin–Ventzell's monograph (Freidlin and Ventzell 2012) (see section 2.7.3).

2.7.1. *Asymptotically small diffusion*

The DMP in discrete–continuous time is given in the series scheme with a small parameter of the series $\varepsilon \to 0$ ($\varepsilon > 0$) by the solution of the following SDE:

$$\Delta_\varepsilon \zeta_\varepsilon(t) = -\varepsilon^3 V(\zeta_\varepsilon(t)) + \varepsilon^2 \sigma \Delta_\varepsilon \mu(t), \quad t = k\varepsilon^3, \quad k \geq 0. \qquad [2.7.1]$$

The discrete–continuous time means the following scaling of increments:

$$\Delta_\varepsilon \alpha(t) := \alpha(t + \varepsilon^3) - \alpha(t), \quad \alpha(t) = \zeta_\varepsilon(t) \text{ or } \mu(t). \qquad [2.7.2]$$

The stochastic component $\mu(t)$, $t \geq 0$, is a martingale difference with the characteristics

$$E\mu(t) = 0, \quad E[\Delta_\varepsilon \mu(t)]^2 = 1. \qquad [2.7.3]$$

The DMPs [2.7.1]–[2.7.3] are given by the following transition probabilities (Korolyuk and Limnios 2005):

$$P_\varepsilon \varphi(c) = E\big[\varphi\big(c + \Delta_\varepsilon \zeta_\varepsilon(t)\big) \mid \zeta_\varepsilon(t) = c\big]. \qquad [2.7.4]$$

The generating operator (generator) of the DMPs [2.7.1]–[2.7.4] has the following expression:

$$Q_\varepsilon \varphi(c) = E\big[\varphi\big(c + \Delta_\varepsilon \zeta_\varepsilon(t)\big) - \varphi(c) \mid \zeta_\varepsilon(t) = c\big]. \qquad [2.7.5]$$

The approximation of the DMPs with ASD, given by SDEs [2.7.1]–[2.7.5], uses the martingale characterization (Korolyuk and Limnios 2005):

$$\mu^\varepsilon(t) = \varphi(\zeta_\varepsilon(t)) - \varphi(\zeta_\varepsilon(0)) - \int_0^{\varepsilon^3[t/\varepsilon^3]} L_\varepsilon \varphi(\zeta_\varepsilon(u))du, \quad t = k\varepsilon^3. \quad [2.7.6]$$

The upper integration limit in [2.7.6] has the asymptotics $\lim_{\varepsilon \to 0} \varepsilon^3[t/\varepsilon^3] = t$.

The normalized generator of the DMP L_ε in the martingale characterization [2.7.6], due to the discrete–continuous time normalization $k = [t/\varepsilon^3]$, is determined by

$$L_\varepsilon \varphi(c) = \varepsilon^{-3} Q_\varepsilon \varphi(c). \qquad [2.7.7]$$

The following theorem establishes that the DMPs [2.7.1]–[2.7.5] are approximated by ASD.

THEOREM 2.7.1.– *The normalized generators* [2.7.7] *have the following asymptotic representation, as* $\varepsilon \to 0$:

$$L_\varepsilon \varphi(c) = L_\varepsilon^0 \varphi(c) + \varepsilon R_\varepsilon \varphi(c), \qquad [2.7.8]$$

by the limit generator of an Ornstein–Uhlenbeck process:

$$L_\varepsilon^0 \varphi(c) = -V(c)\varphi'(c) + \varepsilon \frac{\sigma^2}{2} \varphi''(c), \qquad [2.7.9]$$

and the residual term is neglectable:

$$R_\varepsilon\varphi(c) \to 0, \quad \varepsilon \to 0, \quad \varphi(c) \in C^3(\mathbb{R}). \qquad [2.7.10]$$

Proof of Theorem 2.7.1.– The generating operator

$$Q_\varepsilon\varphi(c) = (P_\varepsilon - \mathbb{I})\varphi(c), \qquad [2.7.11]$$

in virtue of definition [2.7.4] of DMP transition probabilities, admits the following Taylor's decomposition:

$$\begin{aligned} Q_\varepsilon\varphi(c) &= E\big[\Delta_\varepsilon\zeta_\varepsilon(t)\varphi'(c) + \frac{1}{2}\left(\Delta_\varepsilon\zeta_\varepsilon(t)\right)^2\varphi''(c) + \varepsilon^4 R_\varepsilon(c)\varphi\big] \\ &= -\varepsilon^3 V(c)\varphi'(c) + \frac{1}{2}\varepsilon^4\sigma^2\varphi''(c) + \varepsilon^4 R_\varepsilon\varphi(c) \end{aligned} \qquad [2.7.12]$$

on sufficiently smooth test functions $\varphi(c) \in C^3(\mathbb{R})$.

Substituting the decomposition [2.7.12] in [2.7.7], we obtain

$$L_\varepsilon\varphi(c) = -V(c)\varphi'(c) + \frac{1}{2}\varepsilon\sigma^2\varphi''(c) + \varepsilon R_\varepsilon\varphi(c). \qquad [2.7.13]$$

The generator of ASD is determined by the first two components of decomposition [2.7.13]:

$$L_\varepsilon^0\varphi(c) := -V(c)\varphi'(c) + \frac{1}{2}\sqrt{\varepsilon}\sigma^2\varphi''(c). \qquad [2.7.14]$$

The generator L_ε^0 defines an Ornstein–Uhlenbeck-type continuous process $\zeta_\varepsilon^0(t)$, $t \geq 0$, with ASD:

$$d\zeta_\varepsilon^0(t) = -V(\zeta_\varepsilon^0(t)) + \sqrt{\varepsilon}\sigma\, dW(t), \qquad [2.7.15]$$

where the stochastic component is characterized as follows:

$$E[dW(t)] = 0, \quad E[dW(t)]^2 = dt. \qquad [2.7.16]$$

The approximation of the DMP by ASD means the following:

$$\left(L_\varepsilon - L_\varepsilon^0\right)\varphi(c) = \varepsilon R_\varepsilon\varphi(c), \ \ |\,\varepsilon R_\varepsilon\varphi(c)\,| = o(\varepsilon), \quad \varepsilon \to 0. \qquad [2.7.17]$$

The proof of Theorem 2.7.1 is complete. □

Consider the martingale characterization of ASD in the following form:

$$\mu_0^\varepsilon(t) = \varphi(\zeta_\varepsilon^0(t)) - \varphi(\zeta_\varepsilon^0(0)) - \int_0^{\varepsilon^3[t/\varepsilon^3]} L_\varepsilon^0 \varphi(\zeta_\varepsilon^0(u))\, du. \qquad [2.7.18]$$

According to the model limit theorem (Korolyuk and Limnios 2005, Ch. 6, section 3, Theorem 6.3), the following asymptotic relation, in probability, holds:

$$\mu^\varepsilon(t) - \mu_0^\varepsilon(t) \xrightarrow{P} 0, \quad \varepsilon \to 0, \qquad [2.7.19]$$

provided by

$$R_\varepsilon \varphi(c) \to 0, \quad \varepsilon \to 0, \quad \varphi \in C^2(\mathbb{R}). \qquad [2.7.20]$$

The approximation [2.7.15] means an essential extension of the results for Ornstein–Uhlenbeck processes to the class of DMP processes given by the solutions of the SDE [2.7.1]. This is also the novelty of the method.

2.7.2. *EGs of DMP*

The DMP approximation by an Ornstein–Uhlenbeck process with ASD generates an exponential martingale characterization Feng and Kurtz 2006, Ch. 1, section 2.3) with the following scaling:

$$\mu_e^\varepsilon(t) = \exp\Big\{\varphi(\zeta^\varepsilon(t)) - \varphi(\zeta^\varepsilon(0)) - \int_0^{\varepsilon^3[t/\varepsilon^3]} H^\varepsilon \varphi(\zeta^\varepsilon(c))\, ds\Big\}, \; t = k\varepsilon^3. \qquad [2.7.21]$$

The corresponding EG (Borovskikh and Korolyuk 1997, § 2.5) is

$$H^\varepsilon \varphi(c) := \varepsilon^{-2} \ln\Big[1 + e^{-\varphi(c)/\varepsilon}\, Q_\varepsilon\, e^{\varphi(c)/\varepsilon}\Big]. \qquad [2.7.22]$$

EG is generated by the nonlinear exponential semi-group

$$H_t^\varepsilon \varphi(c) = \ln E\Big[e^{\varphi(\zeta^\varepsilon(t))/\varepsilon} \mid \zeta^\varepsilon(0) = c\Big]. \qquad [2.7.23]$$

The generating operator of the DMP [2.7.5] has the form

$$Q_\varepsilon \varphi(c) = E[\varphi(c + \Delta_\varepsilon \zeta^\varepsilon(t)) - \varphi(c) \mid \zeta^\varepsilon(t) = c]. \qquad [2.7.24]$$

In accordance with the method of large deviations analysis for the MPs developed in the monograph (Feng and Kurtz 2006), the first stage of large deviations approach for MP $\zeta^{\varepsilon}(t), t \geq 0$, is solved by the limit theorem (Feng and Kurtz 2006, Ch. 3).

THEOREM 2.7.2.– *The exponential DMP generator* [2.7.22] *is calculated by the passage to limit*

$$\lim_{\varepsilon \to 0} H^{\varepsilon} \varphi(c) = H^0 \varphi(c), \quad \varphi(c) \in \mathcal{C}^3(\mathbb{R}). \qquad [2.7.25]$$

The limit exponential DMP generator [2.7.25] *is determined by*

$$H^0 \varphi(c) = -V(c)\varphi'(c) + 1/2\sigma^2[\varphi'(c)]^2. \qquad [2.7.26]$$

Proof of Theorem 2.7.2.– The main idea is based on the asymptotic relation

$$e^{-\varphi(c)/\varepsilon} Q_{\varepsilon}\, e^{\varphi(c)/\varepsilon} = \varepsilon^2[H^0\varphi(c) + R_{\varepsilon}\varphi(c)]. \qquad [2.7.27]$$

Hereafter, $R_{\varepsilon}\varphi$ means a neglectable term, that is,

$$R_{\varepsilon}\varphi \to 0, \quad \varepsilon \to 0, \quad \varphi \in C^3(R).$$

Justification of formula [2.7.26] is based on the asymptotic representation

$$e^{-\varphi(c)/\varepsilon} Q_{\varepsilon}\, e^{\varphi(c)/\varepsilon} = E[\exp \Delta_{\varepsilon}\varphi(c) - 1 \mid \zeta^{\varepsilon}(t) = c],$$
$$\Delta_{\varepsilon}\varphi(c) := [\varphi(c + \Delta_{\varepsilon}\zeta^{\varepsilon}(t)) - \varphi(c)]/\varepsilon. \qquad [2.7.28]$$

From Taylor's formula,

$$\Delta_{\varepsilon}\varphi(c) = \Delta_{\varepsilon}\zeta^{\varepsilon}(t)\varphi'(c)/\varepsilon + 1/2(\Delta_{\varepsilon}\zeta^{\varepsilon}(t))^2\varphi''(c)/\varepsilon + R_{\varepsilon}\varphi(c). \qquad [2.7.29]$$

The neglectable term $R_{\varepsilon}\varphi(c)$, together with the second term [2.7.29], is in the order of $O(\varepsilon^3)$. Hence, only the first term in [2.7.29] gives the asymptotic contribution.

Using the Taylor expansion up to the second order,

$$e^{\Delta} - 1 = \Delta + 1/2\Delta^2 + O(\Delta^3), \qquad [2.7.30]$$

we obtain the asymptotic representation of formula [2.7.28], computed as follows:

$$
\begin{aligned}
&E[\exp \Delta_\varepsilon \varphi(c) - 1 \mid \zeta^\varepsilon(t) = c] \\
&\quad = E_c[\Delta_\varepsilon \zeta^\varepsilon(t)/\varepsilon]\varphi'(c) + \frac{1}{2} E_c[(\Delta_\varepsilon \zeta^\varepsilon(t)/\varepsilon)^2](\varphi')^2(c) + R_\varepsilon \varphi. \qquad [2.7.31]
\end{aligned}
$$

The first term in the right-hand side of [2.7.31] becomes

$$E_c[\Delta_\varepsilon \zeta^\varepsilon(t)/\varepsilon]\varphi'(c) = -\varepsilon^2 V(c)\varphi'(c) + \varepsilon^3 R_\varepsilon \varphi(c). \qquad [2.7.32]$$

The second term in the right-hand side of [2.7.31] becomes:

$$E_c[(\Delta_\varepsilon \zeta^\varepsilon(t)/\varepsilon)\varphi'(c)]^2 = \varepsilon^2 \sigma^2 (\varphi'(c))^2 + \varepsilon^3 R_\varepsilon \varphi(c). \qquad [2.7.33]$$

So,

$$E_c[e^{\Delta_\varepsilon \varphi(c)} - 1] = \varepsilon^2[-V(c)\varphi'(c) + \frac{1}{2}\sigma^2(\varphi'(c))^2] + \varepsilon^3 R_\varepsilon \varphi(c). \qquad [2.7.34]$$

Finally, we have

$$e^{-\varphi(c)/\varepsilon} Q_\varepsilon \, e^{\varphi(c)/\varepsilon} = \varepsilon^2[H^0 \varphi(c) + \varepsilon R_\varepsilon \varphi(c)], \qquad [2.7.35]$$

where the limit EG is

$$H^0 \varphi(c) := -V(c)\varphi'(c) + \frac{1}{2}\sigma^2(\varphi'(c))^2. \qquad [2.7.36]$$

The proof of Theorem 2.7.2 is complete. □

REMARK 2.7.1.– The EG [2.7.26] sets the martingale (exponential) characterization of the Ornstein–Uhlenbeck process

$$d\zeta^0_\varepsilon(t) = -V(\zeta^0_\varepsilon(t))\,dt + \sqrt{\varepsilon}\sigma\,dW(t), \quad t \geq 0, \qquad [2.7.37]$$

set by the generator [2.7.7].

LEMMA 2.7.1.– *The EG of the Ornstein–Uhlenbeck process* [2.7.37], *set by the generator* [2.7.7]–[2.7.10], *has a limit representation, as* $\varepsilon \to 0$*:*

$$\lim_{\varepsilon \to 0} H^\varepsilon_\zeta \varphi(c) = \lim_{\varepsilon \to 0} e^{-\varphi(c)/\varepsilon} \varepsilon L^0_\varepsilon \, e^{\varphi(c)/\varepsilon} = H^0 \varphi(c), \qquad [2.7.38]$$

where H^0 *is defined in formula* [2.7.26].

REMARK 2.7.2.– The assertions of Theorem 2.7.2 and Lemma 2.7.4 mean that the EG [2.7.26] raises the problem of large deviations for the DMP [2.7.1]–[2.7.3], also for the Ornstein–Uhlenbeck processes [2.7.37].

2.7.3. *AF of DMPs*

In the monograph of Freidlin and Ventzell (2012, Ch. 4), the problem of leaving a region of a dynamical system with Gaussian (asymptotically small) perturbances [2.7.37] is investigated with application of AF, which is given *a priori* by the equality:

$$S_T(\varphi) = \frac{1}{2\sigma^2}\int_0^T [\dot{\varphi}_t + V(\varphi_t)]^2\, \mathrm{d}t. \qquad [2.7.39]$$

Following the monograph, Feng and Kurtz (2006, Part 1), the AF of the Ornstein–Uhlenbeck process [2.7.37], which is characterized by EG [2.7.26], is realized by the solution of the variational problem (Frechet–Legendre transform):

$$\mathcal{L}(c,q) = \sup_p \Big\{pq - H^0(c,p)\Big\}. \qquad [2.7.40]$$

The AF of the MP $\zeta^\varepsilon(t)$, $t \geq 0$, is determined by the integral on test functions $\varphi_t \in C_T^1(\mathbb{R})$, $\dot{\varphi}_t := \mathrm{d}\varphi_t/\mathrm{d}t$:

$$S_T(\varphi) = \int_0^T \mathcal{L}(\varphi_t, \dot{\varphi}_t)\, \mathrm{d}t. \qquad [2.7.41]$$

The function of two variables $H^0(c,p)$ is determined by the limit EG [2.7.26]:

$$H^0(c,\dot{\varphi}) \equiv H^0\varphi(c). \qquad [2.7.42]$$

Therefore,

$$H^0(c,\dot{\varphi}) = -V(c)\dot{\varphi} + 1/2\sigma^2\dot{\varphi}^2. \qquad [2.7.43]$$

The fact of coincidence of the EG of the Ornstein–Uhlenbeck process with ASD [2.7.15] and [2.7.16] with the EG of the DMP [2.7.21]–[2.7.25] results in the following theorem.

THEOREM 2.7.3.– *The AFs of the Ornstein–Uhlenbeck process* [2.7.37] *and for the DMP* [2.7.1]–[2.7.3] *coincide and are determined by the integral* [2.7.39].

PROOF 2.33.– The variational problem [2.7.40], taking into account [2.7.41], involves the Frechet–Legendre transformation in [2.7.40] for the integral function [2.7.41]:

$$\mathcal{L}(c,q) = [q + V(c)]^2. \qquad [2.7.44]$$

Substituting the test function $\varphi_t = c$, $\dot{\varphi}_t = q$, we obtain the assertion of Theorem 2.7.3. □

The discrete Markov diffusion, generated by the solutions of the SDE [2.7.1], approximates the real SEs with normalization (Koroliouk 2015c):

$$\zeta^{\varepsilon}(t) = S_N(k) - \rho, \quad t = k\varepsilon^3, \quad k \geq 0. \qquad [2.7.45]$$

Given the values range of real SEs $0 \leq S_N(k) \leq 1$, we determine the sojourn interval for the approximating SEs as follows:

$$(-\rho,\, 1-\rho), \quad 0 \leq \rho \leq 1. \qquad [2.7.46]$$

Now we have the opportunity to use the results of Freidlin–Ventzell's monograph adapted to our problem (a variant of Theorem 1.2; Freidlin and Ventzell 2012, Ch. 4).

THEOREM 2.7.4.– *The probability of exit the SEs* [2.7.45] *and* [2.7.46] *from interval* $(-\rho, 1-\rho)$ *is given by the following asymptotic relation (Koroliouk 2016c):*

$$\lim_{\varepsilon\to 0} \varepsilon^2 \ln P\{\zeta^{\varepsilon}(t) \in (-\rho, 1-\rho) \mid \zeta^{\varepsilon}(0) = c\} = -\min_{\varphi\in\overline{H}(c)} S_T(\varphi), \quad [2.7.47]$$

in which

$$\overline{H}(c) := \{\varphi_t : \varphi_0 = c,\ \varphi_t \in [-\rho, 1-\rho], \quad 0 \leq t \leq T\}.$$

In addition, the following asymptotic relation holds:

$$\lim_{\varepsilon\to 0} \varepsilon^2 \ln P\{\tau^{\varepsilon} \leq t\} = -\min_{\varphi\in H(c)} S_T(\varphi), \qquad [2.7.48]$$

for the exit time of the interval:

$$\tau^{\varepsilon} := \min_{\varphi\in H(c)} \{t : \zeta^{\varepsilon}(t) \notin (-\rho, 1-\rho)\}.$$

Here, by definition, the set of test functions is,

$$H(c) := \{\varphi_t : \varphi_0 = c,\ \varphi_t \notin (-\rho, 1-\rho), \quad 0 \leq t \leq T\}.$$

For a dynamic system with the linear velocity

$$dS(t)/dt = -VS(t), \quad -\infty \leq t \leq T, \tag{2.7.49}$$

the potential $U(c)$ is determined by the relation (Freidlin and Ventzell 2012, §4.2, c. 149)

$$U(c) = 1/2Vs^2, \quad s \in R. \tag{2.7.50}$$

Denote the solution of the dynamic system

$$d\widehat{S}(t)/dt = V\widehat{S}(t), \quad \widehat{S}(-\infty) = 0, \quad \widehat{S}(T) = c, \quad -\infty \leq t \leq T. \tag{2.7.51}$$

The following equality is obvious:

$$\widehat{S}_0(t) = ce^{V(t-T)}. \tag{2.7.52}$$

This is a variant of Theorem 2.8.1 Freidlin and Ventzell 2012, Section 4.3, p. 162).

THEOREM 2.7.5.– *The DMPs* [2.7.45] *and* [2.7.46] *are characterized by the potential* [2.7.50] *of the dynamic system* [2.7.49], *namely, there is the inequality*

$$S_T(\varphi_t) \geq \frac{1}{\sigma^2} V s^2 \quad \forall \{\varphi_t : \varphi_0 = 0, \ \varphi_T = c\}. \tag{2.7.53}$$

PROOF 2.34.– The inequality [2.7.53] follows from the obvious identity

$$S_T(\varphi_t) = \frac{1}{\sigma^2} \left[\frac{1}{2} \int_0^T [\dot{\varphi}_t - V\varphi_t]^2 \, dt + Vs^2 \right]. \tag{2.7.54}$$

So, for the test functions $\varphi_t : \varphi_0 = 0$ and $\varphi_T = c$, the inequality [2.7.53] takes place. □

REMARK 2.7.3.– Given the solution [2.7.51] of the dynamic system [2.7.49], we can also argue the validity of the inequality

$$S_T(\varphi_t) \geq \widehat{S}_T(\widehat{\varphi}_t), \tag{2.7.55}$$

where the test function $\widehat{\varphi}_t$ is determined by the solution of the dynamic system $\widehat{\varphi}_t$:

$$\dot{\widehat{\varphi}}_t - V\widehat{\varphi}_t = 0, \quad \widehat{\varphi}_{-\infty} = 0, \quad \widehat{\varphi}_T = c. \tag{2.7.56}$$

So, the unique extremum of AF [2.7.41] is given by the solution of the dynamic system [2.7.56], namely

$$\widehat{\varphi}_t = ce^{V(t-T)}, \quad -\infty \leq t \leq T. \qquad [2.7.57]$$

So, the inequality [2.7.55] is established.

Asymptotic analysis of exit from the interval $(-\rho, 1-\rho)$ of SEs [2.7.21]–[2.7.25] can be realized using approximation of the process $\zeta^\varepsilon(t)$, $t \geq 0$, by the Ornstein–Uhlenbeck process $\zeta^0_\varepsilon(t)$, $t \geq 0$, with the generator [2.7.14].

Consider the distribution functions of the Ornstein–Uhlenbeck process (see Freidlin and Ventzell 2012, section 4.1):

$$U^\varepsilon_t(c) = P\{\zeta^0_\varepsilon(t) \in (-\rho, 1-\rho) \mid \zeta^0_\varepsilon(0) = c\}, \qquad [2.7.58]$$

$$\overline{U}^\varepsilon_t(c) = P\{\tau^\varepsilon \leq t) \mid \zeta^0_\varepsilon(0) = c\}. \qquad [2.7.59]$$

Using Theorem 3.1 (Freidlin and Ventzell 2012, section 4.3, p. 162), we can refine the results of Theorem 1.2 (Freidlin and Ventzell 2012, section 4.1, p. 145).

We can make sure that the distribution function [2.7.58] is determined by the solution of a boundary problem for a differential equation of second order with a small parameter ε in the term of the highest derivative (see, for example, Freidlin and Ventzell 2012, Ch. 4):

$$\partial U^\varepsilon_t(c)/\partial t = L^0_\varepsilon U^\varepsilon_t(c), \quad U^\varepsilon_0(c) = 1, \ s \in (-\rho, 1-\rho),$$
$$U^\varepsilon_0(c_0) = 0, \ s_0 \in \{-\rho, 1-\rho\}, \qquad [2.7.60]$$

In particular, the probability of leaving the interval (ρ_-, ρ_+), $\rho_- := -\rho$, $\rho_+ := 1-\rho$:

$$U^\varepsilon_\pm(c) := P\{\zeta^0_\varepsilon(\tau^\varepsilon) = \rho_\pm \mid \zeta^0_\varepsilon(0) = c\}, \quad \rho_- < s < \rho_+ \qquad [2.7.61]$$

is determined by the solution of the boundary elliptic problem

$$L^0_\varepsilon U^\varepsilon_\pm(c) = 0, \quad U^\varepsilon_\pm(\rho_\pm) = 1, \quad U^\varepsilon_\pm(\rho_\mp) = 0. \qquad [2.7.62]$$

The solution of the boundary problem [2.7.62] is given by

$$U^\varepsilon_+(c) = \int_{\rho_-}^{s} \frac{\exp[(V/\varepsilon\sigma^2)r^2]dr}{\bar{U}_\varepsilon}, \qquad [2.7.63]$$

$$U_-^\varepsilon(c) = \int_s^{\rho_+} \frac{\exp[(V/\varepsilon\sigma^2)r^2]\mathrm{d}r}{\bar{U}_\varepsilon}. \quad [2.7.64]$$

Here the normalizing constant

$$\bar{U}_\varepsilon = \int_{\rho_-}^{\rho_+} \exp\left[\frac{V}{\varepsilon\sigma^2}r^2\right] \mathrm{d}r. \quad [2.7.65]$$

COROLLARY 2.7.1.– *The probabilities of leaving the interval* (ρ_-, ρ_+) *are characterized by the limit behavior*

$$\lim_{\varepsilon\to 0} P\{\zeta_\varepsilon^0(\tau^\varepsilon) = \rho_+ \mid \zeta_\varepsilon^0(0) = c\} = \begin{cases} 1, & \rho > 1/2, \\ 0, & \rho < 1/2. \end{cases} \quad [2.7.66]$$

In particular, we can see that there is the asymptotic probability behavior [2.7.60] *as* $\rho = 1/2$*:*

$$\lim_{\varepsilon\to 0} P\{\zeta_\varepsilon^0(\tau^\varepsilon) = \pm\frac{1}{2} \mid \zeta_\varepsilon^0(0) = c\} = 1/2. \quad [2.7.67]$$

3

Statistics of Statistical Experiments

In this chapter, two main problems of mathematical statistics are examined: parameter estimation and hypothesis verification for the dynamic processes of statistical experiments (SEs) as $k \to \infty$.

The *statistical estimation* of the drift parameter V_0 under additional conditions of SE wide-sense stationarity, is carried out in terms of covariances of two-component processes $\{\alpha_t, \Delta\alpha_{t+1}\}$, $t \geq 0$.

The *hypothesis verification* for the dynamic SE process, as $k \to \infty$, is reduced to SE classification, determined by the values of the drift parameter V_0 and the equilibria $\rho_\pm$.

3.1. Parameter estimation of one-dimensional stationary SEs

3.1.1. *Stationarity*

As set out in Chapter 1 (see Proposition 1.1.3), the stationary binary SEs are determined by the solutions of the *stochastic difference equations* (SDEs).

DEFINITION 3.1.1.– An SE is determined by the sequence α_t, $t \in \mathcal{N}_0 := \{0, 1, 2, 3, ...\}$, the increments of which

$$\Delta\alpha_{t+1} := \alpha_{t+1} - \alpha_t, \quad t \geq 0, \quad \alpha_0 \text{ is given},$$

are defined by the following SDE:

$$\Delta\alpha_{t+1} = -V_0\alpha_t + \Delta\mu_{t+1}, \quad t \geq 0. \qquad [3.1.1]$$

The predicted component is a linear function of SEs with a drift parameter $V_0 > 0$. The stochastic components $\Delta\mu_{t+1}, t \geq 0$, are given by the martingale differences

$$\Delta\mu_{t+1} := \mu_{t+1} - \mu_t, \quad t \geq 0, \quad \mu_0 = 0,$$

which generate a stationary square-integrated martingale

$$\mu_t := \sum_{k=0}^{t-1} \Delta\mu_{k+1}, \quad t \in \mathcal{N}_0,$$

characterized by the first two moments

$$E[\Delta\mu_{t+1}] = 0, \quad E[(\Delta\mu_{t+1})^2] =: \sigma^2, \quad t \geq 0. \qquad [3.1.2]$$

Let us introduce the denotation of covariances

$$E\alpha_t^2 = \sigma_0^2, \quad E[\alpha_t \alpha_{t+s}] =: R(s), \quad t, s \geq 0. \qquad [3.1.3]$$

The wide-sense stationarity of the SEs α_t, $t \geq 0$, implies uncorrelatedness with the stochastic component: $E[\alpha_t \, \Delta\mu_{t+1}] = 0$, as well as the stationarity of an initial value α_0, which is recalled in the following theorem.

THEOREM 3.1.1.– (Stationarity theorem). *SEs* [3.1.1]–[3.1.3] *are wide-sense stationary, random sequence if and only if the following relations hold:*

$$E\alpha_0 = 0, \quad \sigma^2 = \sigma_0^2 \, V_0 \, (2 - V_0). \qquad [3.1.4]$$

Proof of Theorem 3.1.1.– Conducted by induction.

First of all, the uncorrelation of the SEs with the martingale differences is used

$$E(\alpha_t \cdot \Delta\mu_{t+1}) = 0.$$

Let $B_t := E\alpha_t^2 = \sigma_0^2$, $t \leq T$, then we have the following relations:

$$E[\alpha_t \alpha_{t+1}] = E[\alpha_t(\alpha_t + \Delta\alpha_{t+1})] = B_t + R_\alpha^0, \quad R_\alpha^0 := E[\alpha_t \, \Delta\alpha_{t+1}],$$

$$R_\alpha^0 = E[\alpha_t(-V_0\alpha_t + \sigma\Delta\mu_{t+1})] = -V_0 B_t. \qquad [3.1.5]$$

Note that

$$E[\Delta\alpha_{t+1}]^2 = E[-V_0\alpha_t + \sigma\Delta\mu_{t+1}]^2 = V_0^2 B_t + \sigma^2,$$
$$E[\alpha_{t+1} - \alpha_t]^2 = B_{t+1} - B_t - 2R_\alpha^0. \qquad [3.1.6]$$

Equating the right-hand sides of equations [3.1.6],

$$B_{t+1} - B_t = 2R_\alpha^0 + V_0^2 B_t + \sigma^2.$$

Using [3.1.5],

$$B_{t+1} - B_t = -2V_0 B_t + V_0^2 B_t + \sigma^2 = -(2V_0 - V_0^2)B_t + \sigma^2. \qquad [3.1.7]$$

Taking into account [3.1.7] and the stationarity condition $B_{t+1} - B_t = 0$,

$$\sigma^2 = (2V_0 - V_0^2)\sigma_0^2.$$

So, for everyone, $t \in \mathcal{N}_0$, $B_t = \sigma_0^2$.

The necessity of condition [3.1.4] is obvious. □

The solution of the SDEs [3.1.1] and [3.1.2] and Theorem 3.1.1 provides the wide-sense stationarity of SEs α_t, $t \geq 0$, with the following covariance function:

$$R(s) := \text{cov}(\alpha_{t+s}, \alpha_t) = E[\alpha_{t+s}\alpha_t] = \sigma_0^2 q^s, \quad q := 1 - V. \qquad [3.1.8]$$

REMARK 3.1.1.– The correctness of the stationarity condition [3.1.4] is provided by an additional parameter limitation:

$$0 < V_0 < 2.$$

The wide-sense stationary, SEs α_t, $t \geq 0$, are characterized by a matrix of covariances of a two-component vector $(\alpha_t, \Delta\alpha_{t+1})$, $t \geq 0$.

THEOREM 3.1.2.– *The covariance matrix*

$$\mathbb{R} = \begin{bmatrix} R & R^0 \\ R^0 & R^\Delta \end{bmatrix} \qquad [3.1.9]$$

of the two-component vector $(\alpha_t, \Delta\alpha_t)$, $t \geq 0$, *is given by the relationships:*

$$R := E\alpha_t^2 = \sigma_0^2, \quad R^\Delta := E[\Delta\alpha_{t+1}]^2 = 2V_0\sigma_0^2,$$
$$R^0 := E[\alpha_t\Delta\alpha_{t+1}] = -V_0\sigma_0^2. \qquad [3.1.10]$$

Proof of Theorem 3.1.2.– The components of the covariance matrix [3.1.10] are calculated using the SDE [3.1.1], the stationarity conditions [3.1.4] and also the uncorrelation of the stochastic components $\Delta\mu_{t+1}$ and α_t. So, the relations

$$\begin{aligned} R^{\Delta} &= -V_0\sigma_0^2 + \sigma^2 = V_0\sigma_0^2 + \sigma_0^2(2V_0 - V_0^2) = 2V_0\sigma_0^2, \\ R^0 &= -V_0\sigma_0^2 + E[\alpha_t \cdot \Delta\mu_{t+1}] = -V_0\sigma_0^2, \end{aligned} \qquad [3.1.11]$$

prove expressions [3.1.10]. □

An important additional assumption on the stationary SEs is Gaussian distribution of the stochastic component, that is, the SDE [3.1.1] has the form

$$\Delta\alpha_{t+1} = -V_0\alpha_t + \Delta W_{t+1}, \quad t \in \mathcal{N}_0. \qquad [3.1.12]$$

The stochastic component ΔW_{t+1}, $t \in \mathcal{N}_0$, is given by a sequence of normally distributed and jointly independent random variables, characterized by two first moments:

$$E(\Delta W_{t+1}) = 0, \quad E(\Delta W_{t+1})^2 = \sigma^2, \quad t \in \mathcal{N}_0. \qquad [3.1.13]$$

The stationarity condition [3.1.4] now additionally implies the normal distribution of the initial value α_0 with the first two moments

$$\begin{aligned} &E\alpha_0 = 0, \quad E\alpha_0^2 = \sigma_0^2 = \tfrac{\sigma^2}{\mathcal{E}_0}, \\ &\mathcal{E}_0 := 2V_0\left(1 - \tfrac{V_0}{2}\right). \end{aligned} \qquad [3.1.14]$$

The parameter $\mathcal{E}_0$ is called the *stationarity factor.*

COROLLARY 3.1.1.– *The covariance matrix of a two-dimensional stationary Markov Gaussian sequence* $(\alpha_t, \Delta\alpha_{t+1})$ *has the form*

$$\mathbb{R} = \sigma_0^2 \begin{bmatrix} 1 & -V_0 \\ -V_0 & 2V_0 \end{bmatrix} \qquad [3.1.15]$$

and generates a quadratic form of two-dimensional normal distribution of the vector $(\alpha_t, \Delta\alpha_{t+1})$*, for any* $t \in \mathbb{N}_0$*:*

$$\mathbb{R}^{-1} = \frac{1}{\sigma^2} \begin{bmatrix} 2V_0 & V_0 \\ V_0 & 1 \end{bmatrix}. \qquad [3.1.16]$$

The easiest way to check the validity of formula [3.1.16] *is to multiply the matrices* [3.1.15] *and* [3.1.16], *taking into account the stationarity conditions* [3.1.4].

COROLLARY 3.1.2.– *The matrix of quadratic form* [3.1.16], *which generates two-dimensional normal distribution of the vector* $(\alpha_t, \Delta\alpha_{t+1})$, *for each* $t \in \mathbb{N}_0$, *can be represented by elements of the covariance matrix* [3.1.15] *(see Gnedenko 2011, sections 4:20 and 5:24):*

$$\frac{1}{R_{11}^{(-1)}} = WR, \quad \frac{1}{R_{22}^{(-1)}} = WR^{\Delta}, \qquad [3.1.17]$$

$$\frac{1}{R_{12}^{(-1)}} = \frac{1}{R_{21}^{(-1)}} = -\left(\frac{2}{V_0}\right) WR^0. \qquad [3.1.18]$$

Here by definition

$$W := 1 - \frac{V_0}{2}. \qquad [3.1.19]$$

The validity of formulae [3.1.17] and [3.1.18] follows from the relation

$$\sigma^2 = \sigma_0^2 \cdot 2V_0W. \qquad [3.1.20]$$

COROLLARY 3.1.3.– *Formulae* [3.1.10] *generate covariance representations of the drift parameter* V_0*:*

$$V_0 = -\frac{R^0}{R} \qquad [3.1.21]$$

and

$$V_0 = \frac{R^{\Delta}}{2R}. \qquad [3.1.22]$$

COROLLARY 3.1.4.– *Formulae* [3.1.21] *and* [3.1.22] *also determine the correlation coefficient*

$$r^2 = \frac{(R^0)^2}{RR^{\Delta}} = \frac{V_0}{2}. \qquad [3.1.23]$$

REMARK 3.1.2.– Formulae [3.1.19] and [3.1.23] explain the limitations of the parameter values $0 < V_0 < 2$.

The extreme value $V_0 = 0$ corresponds to the situation in which the predicted component is absent, and the SE turns into a purely stochastic sequence (Moklyachuk 1994).

Another extreme value $V_0 = 2$ means $r = \pm 1$, then the variables α and $\Delta\alpha$ are linearly dependent (Gnedenko 2011, 5:24, section 131). That is, the SE is transformed into a purely deterministic process.

3.1.2. *Covariance statistics*

The trajectories of SEs generate covariance statistics in accordance with the definition of covariances [3.1.10]:

$$R_T := \tfrac{1}{T}\sum_{t=0}^{T-1}\alpha_t^2,\ R_T^0 := \tfrac{1}{T}\sum_{t=0}^{T-1}\alpha_t\Delta\alpha_{t+1},$$
$$R_T^\Delta := \tfrac{1}{T}\sum_{t=0}^{T-1}(\Delta\alpha_{t+1})^2. \qquad [3.1.24]$$

The stationarity of process α_t, $t \geq 0$, provides unbiased estimations [3.1.24]:

$$ER_T = R, \quad ER_T^0 = R^0, \quad ER_T^\Delta = R^\Delta. \qquad [3.1.25]$$

The convergence (as $T \to \infty$) of statistics [3.1.24] sets the condition on the fourth moments. For example, the consistency of averaged quadratic statistic R_T is provided by the condition (Shiryaev 2011, Ch. 6, section 4):

$$\frac{1}{T}\sum_{t=0}^{T-1} E\left[(\alpha_t^2 - R)(\alpha_0^2 - R)\right] \to 0\,, \quad T \to \infty. \qquad [3.1.26]$$

The wide-sense stationarity of the process α_t, $t \geq 0$, suggests the existence of only the first two moments and therefore the fulfillment of the condition [3.1.26] is problematic.

Under the additional condition of normality of the stochastic component [3.1.12] and [3.1.13], Corollary 3.1.2 allows us to specify the quadratic form of a two-component Gaussian vector $(\alpha_t, \Delta\alpha_{t+1})$, with matrix [3.1.16], in the following simplified form:

$$R^{-1}(\alpha,\, \Delta\alpha) = \frac{2V_0(\alpha^2 + \alpha\Delta\alpha) + (\Delta\alpha)^2}{WR^\Delta}. \qquad [3.1.27]$$

Using the relationship of the drift parameter V_0 with the coefficient of correlation r:

$$W = 1 - r^2, \qquad [3.1.28]$$

where the correlation r^2 is defined in [3.1.23], as well as the identity

$$RR^{\Delta} = 2V_0R^2, \qquad [3.1.29]$$

we can verify that the quadratic form [3.1.27] is transformed into a standard form with elements [3.1.16].

Now the normal distribution of stationary process allows us to formulate the consistency condition in terms of covariance statistics [3.1.24].

PROPOSITION 3.1.1.– The covariance statistics [3.1.24] for the stationary Gaussian process, determined by SDE [3.1.12], are *square mean consistent*.

$$E[R_T - R]^2 \to 0, \;\; E[R_T^{\Delta} - R^{\Delta}]^2 \to 0, \;\; E[R_T^0 - R^0]^2 \to 0, \quad T \to \infty. \qquad [3.1.30]$$

Proof of Proposition 3.1.1.– The proof is based on the statement (Bulinski and Shiryaev 2003, section VII, ex. 23) that the necessary and sufficient condition for the statistics R_T, consistency is the convergence of the following covariance function:

$$\frac{1}{T}\sum_{s=0}^{T-1} R^2(s) \xrightarrow{L^2} 0, \quad T \to \infty. \qquad [3.1.31]$$

By the form of covariance function [3.1.8], the necessary and sufficient condition [3.1.31] is evidently fulfilled. Similarly, condition [3.1.31] holds for the statistics R_T^0 and R_T^{Δ}.

Now it remains to convince ourselves that the necessary and sufficient condition [3.1.26] in Gaussian case reduces to condition [3.1.31].

To do this, we use the fact that all cumulants of order $n \geq 3$ of normal distribution are zero. In particular, we compute condition [3.1.26] in the Gaussian case:

$$E\left[(\alpha_t^2 - R)(\alpha_0^2 - R)\right] = E(\alpha_t\alpha_0)^2 - R^2. \qquad [3.1.32]$$

Now, we use the representation of the fourth moments:

$$E(\alpha_t\alpha_0)^2 = R^2 + 2R^2(t). \qquad [3.1.33]$$

Remark. Representation [3.1.33] is a partial case of the general formula of the fourth moments of normal distribution (Shiryaev 2011, II:12, ex. 7):

$$R^{(4)} = E\left[\alpha_0\alpha_{t1}\alpha_{t2}\alpha_{t3}\right] = E\left[(\alpha_0\alpha_{t1}\right] E\left[\alpha_{t2}\alpha_{t3}\right] \\ + E\left[(\alpha_0\alpha_{t2}\right] E\left[\alpha_{t1}\alpha_{t3}\right] + E\left[(\alpha_0\alpha_{t3}\right] E\left[\alpha_{t1}\alpha_{t2}\right]. \qquad [3.1.34]$$

Now the necessary and sufficient condition for the covariance statistics mean-square convergence, taking into account [3.1.26], [3.1.32] and [3.1.33], is reduced to condition [3.1.31]. □

3.1.3. *A priori statistics*

The basis of covariances [3.1.10] with an active (but unknown) drift parameter V_0 gives the reason to introduce *a priori* statistics (Koroliouk 2016b), which estimate the parameter V_0.

PROPOSITION 3.1.2.– There are two *a priori* estimates of the parameter V_0:

$$V_0 \approx V_T^0 := -\frac{R_T^0}{R_T}, \quad V_0 \approx V_T^\Delta := \frac{R_T^\Delta}{2R_T}, \qquad [3.1.35]$$

as well as *a priori* estimation of the parameter σ^2:

$$\sigma^2 \approx B_T^0 := \frac{1}{T}\sum_{t=0}^{T-1}(\Delta\mu_{t+1})^2.$$

The strong law of large numbers for the mean square integrable martingales (Shiryaev 2011, Ch. 7, section 5, Theorem 4) provides the strong consistence of the first *a priori* statistics in [3.1.35].

THEOREM 3.1.3.– *A convergence with probability 1 takes place,*

$$V_T^0 \xrightarrow{P1} V_0, \quad T \to \infty. \qquad [3.1.36]$$

Proof of Theorem 3.1.3.– Let us multiply the SDE [3.1.1] by α_t:

$$\alpha_t\Delta\alpha_{t+1} = -V_0\alpha_t^2 + \Delta\mu_{t+1}\alpha_t \qquad [3.1.37]$$

and summarize [3.1.37] by $t \in [0, T-1]$. Taking into account the definitions [3.1.24], we obtain

$$R_T^0 = -V_0 R_T + \frac{M_T}{T}. \qquad [3.1.38]$$

The second component in the right-hand side is the transformation of martingale (Shiryaev 2011, Ch. 7, section 1):

$$M_T := \sum_{t=0}^{T-1} \alpha_t \Delta\mu_{t+1}. \qquad [3.1.39]$$

Its quadratic characteristic is

$$\langle M\rangle_T = \sum_{t=0}^{T-1} E[\alpha_t^2(\Delta\mu_{t+1})^2 \mid \alpha_t] = \sum_{t=0}^{T-1} \alpha_t^2 E[(\Delta\mu_{t+1})^2 \mid \alpha_t]$$

$$= \sigma^2 \sum_{t=0}^{T-1} \alpha_t^2 = T\sigma^2 R_T. \qquad [3.1.40]$$

Using the definition of V_T^0 given in [3.1.35] and the quadratic characteristic of martingale [3.1.40], we have the following representation for the difference:

$$V_0 - V_T^0 = \frac{\sigma^2 M_T}{\langle M\rangle_T}. \qquad [3.1.41]$$

Now the strong consistency of the first *a priori* estimate in [3.1.35] follows from the condition (Shiryaev 2011, Ch. 7, section 5, Theorem 4):

$$\langle M\rangle_T \xrightarrow{P1} \infty, \quad T \to \infty. \qquad [3.1.42]$$

The proof of Theorem 3.1.3 is complete. □

3.1.4. *Optimal estimating function*

The main statistical problem in the analysis of stationary SEs [3.1.1]–[3.1.4] is to construct a statistical estimation of the drift parameter V_0, which determines the predicted component of the SDE [3.1.1], taking into account the stationarity properties of a two-component vector $(\alpha_t, \Delta\alpha_t)$, $t \geq 0$, characterized by the matrix of covariances [3.1.9].

The problem of statistical estimation, above all, is the choice of the cost function and the corresponding function of risk (Ibragimov and Has'minskii 1981, Ch. 1). As a rule, the function of costs is determined by a quadratic function. Then, the risk function determines the mean-square deviation of the parameter estimation.

The cost function is defined, using the martingale differences

$$M_T(V) := \frac{1}{T}\sum_{t=0}^{T-1} \Delta\mu_t, \quad 0 \le t \le T, \qquad [3.1.43]$$

or in equivalent form

$$M_T(V) = \frac{1}{T}\sum_{t=0}^{T-1} (\Delta\alpha_t + V\alpha_t), \quad 0 \le t \le T. \qquad [3.1.44]$$

DEFINITION 3.1.2.– *Optimal estimating function* (OEF) (Koroliouk 2016b) is defined, using the quadratic variation of the martingale differences [3.1.44]:

$$[M(V)]_T := \frac{1}{T}\sum_{t=0}^{T-1} [\Delta\alpha_t + V\alpha_t]^2. \qquad [3.1.45]$$

The motivation for choosing an OEF in Definition 3.1.2 becomes clear when compared with the estimates for maximum likelihood for Gaussian SEs (section 3.1.5).

Note that

$$[M(V)]_T = V^2 R_T + 2V R_T^0 + R_T^\Delta. \qquad [3.1.46]$$

REMARK 3.1.3.– Let us introduce the normalized OEF and its stochastic equivalent expression, in virtue of [3.1.35]:

$$\frac{[M(V)]_T}{R_T} = V^2 - 2VV_T^0 + 2V_T^\Delta \approx V^2 - 2VV_0 + 2V_0. \qquad [3.1.47]$$

REMARK 3.1.4.– *Quasi-score functional* (QSF) (Heyde 1997, Ch. 2), which characterizes the function of risk, is calculated by the ratio of mathematical expectations:

$$\mathcal{E}(V, V_0) := \frac{E[M_T(V)]}{E[R_T]} = V^2 - 2VV_0 + 2V_0,$$

or equivalently

$$\mathcal{E}(V, V_0) = (V - V_0)^2 + \mathcal{E}_0, \quad \mathcal{E}_0 := 2V_0 - V_0^2. \qquad [3.1.48]$$

REMARK 3.1.5.– Thanks to the wide-sense stationarity of SEs, the normalized QSF $\mathcal{E}(V, V_0)$ does not depend on the interval of observation T.

COROLLARY 3.1.5.– *The normalized QSF is a positively defined function of the parameter V and admits the following representation:*

$$\mathcal{E}(V, V_0) = (V - V_0)^2 + \mathcal{E}_0, \quad \mathcal{E}_0 := \frac{\sigma^2}{\sigma_0^2} = V_0(2 - V_0). \qquad [3.1.49]$$

The stationarity coefficient $\mathcal{E}_0$ plays an important role in optimal estimation.

REMARK 3.1.6.– The OEF [3.1.47] depends on the current value of the drift parameter V $(0 < V < 2)$, as well as on its true value V_0, which defines the paths of the solution of the SDE [3.1.1]. The statistics R_T does not depend on the current value of the reference parameter V.

In the following theorem, the statistics V_T^0 of the drift parameter V_0 is determined by the minimum OEF, which coincides with the minimum of the normalized OEF [3.1.46].

THEOREM 3.1.4.– *A priori statistics V_T^0 of the drift parameter V_0 is determined by the relation of optimality:*

$$\min_{0<V<2} M_T(V, V_0) = \min_{0<V<2} \mathcal{E}_T(V, V_0) = \mathcal{E}_T^0 = 2V_T^\Delta - (V_T^0)^2 \qquad [3.1.50]$$

and given by a priori statistics [3.1.35]*:*

$$V_T^0 = -\frac{R_T^0}{R_T} = -\frac{\sum_{t=1}^T \alpha_t \Delta\alpha_t}{\sum_{t=1}^T \alpha_t^2}, \qquad [3.1.51]$$

$$V_T^\Delta := \frac{R_T^\Delta}{2R_T} = \frac{\sum_{t=1}^T (\Delta\alpha_t)^2}{2\sum_{t=1}^T \alpha_t^2}. \qquad [3.1.52]$$

Proof of Theorem 3.1.4.– The normalized OEF

$$\mathcal{E}_T(V, V_0) = V^2 - 2VV_T^0 + 2V_T^\Delta \qquad [3.1.53]$$

decomposes into two components:

$$\mathcal{E}_T(V, V_0) = (V - V_T^0)^2 + \mathcal{E}_T^0, \quad \mathcal{E}_T^0 := 2V_T^\Delta - (V_T^0)^2, \qquad [3.1.54]$$

where the statistics V_T^0 and V_T^Δ, defined in [3.1.51] and [3.1.52], respectively, depend only on the true value of the parameter V_0 and do not depend on the parameter V.

The last relationship completes the proof of Theorem 3.1.4. □

The residual term $\mathcal{E}_T^0$ in formula [3.1.53] serves as an estimate of the stationarity factor

$$\mathcal{E}_0 := \frac{\sigma^2}{\sigma_0^2} = V_0(2 - V_0). \qquad [3.1.55]$$

Under the mean-square convergence condition of covariance statistics [3.1.30], the stationarity factor statistics also converge in mean-square:

$$E[\mathcal{E}_T^0 - \mathcal{E}_0]^2 \to 0, \quad T \to \infty. \qquad [3.1.56]$$

3.1.5. *Stationary Gaussian SEs*

The stationary Gaussian SE is given by the solution of the SDEs

$$\Delta\alpha_{t+1} = -V_0\alpha_t + \Delta W_{t+1}, \quad t \in \mathcal{N}_0, \quad \mathcal{N}_0 := 0, 1, 2, \ldots. \qquad [3.1.57]$$

The stochastic component ΔW_t is given by a sequence of Gaussian martingale differences W_t, $t \in \mathcal{N}_0$ (Borovkov 1988, Ch. 2, section 6):

$$\Delta W_{t+1} := W_{t+1} - W_t, \quad t \in \mathcal{N}_0, \quad W_0 = 0,$$

which are characterized by the first two moments:

$$E(\Delta W_{t+1}) = 0, \quad E(\Delta W_{t+1})^2 = \sigma^2, \quad t \in \mathcal{N}_0. \qquad [3.1.58]$$

The initial value of SEs [3.1.57] and [3.1.58] is a normally distributed random variable, independent of ΔW_t, $t \geq 1$. The drift parameter V_0 satisfies the constraint $0 < V_0 < 2$.

Optimal estimation of the drift parameter V_0 is constructed using the likelihood function for a stationary Gaussian sequence W_t, $t \in \mathcal{N}_0$, based on the representation

$$\ln L(W_T) = -\frac{1}{2\sigma^2}\frac{1}{T}\sum_{t=0}^{T-1}(\Delta W_{t+1})^2 + P_{0T}, \qquad [3.1.59]$$

where P_{0T} is a constant that does not depend on V.

Now the normalized likelihood function for the predicted parameter V_0 can be represented by relation [3.1.44]:

$$\Delta W_{t+1} = \Delta\alpha_{t+1} + V\alpha_t, \quad t \in \mathcal{N}_0, \qquad [3.1.60]$$

in the following form:

$$\begin{aligned}\mathcal{E}_T(V, V_0) &= \frac{1}{T}\sum_{t=0}^{T-1}\frac{(\Delta W_{t+1})^2}{R_T} = V^2 - 2VV_T^0 + 2V_T^\Delta \\ &= (V - V_T^0)^2 + 2V_T^\Delta - (V_T^0)^2. \qquad [3.1.61]\end{aligned}$$

According to Theorem 3.1.4, the optimal statistic V_T^0 for the drift parameter V_0 is given by formula [3.1.51].

REMARK 3.1.7.– The optimal statistic V_T^0, determined in [3.1.51], for the normalized estimator function [3.1.47], coincides with the estimation of maximum likelihood expressed by a logarithm of the likelihood function [3.1.59].

3.2. Parameter estimators for multivariate stationary SEs

The statistics of one-dimensional stationary SEs, investigated in section 3.1, is generalized on the MSE (Koroliouk 2016a).

In this section, vector stochastic difference equations are used, in order to formulate the wide-sense stationarity conditions of MSEs (Theorem 3.2.1).

Stationary SEs are completely determined by covariance matrices of a two-component process $(\alpha_t, \Delta\alpha_{t+1})$. For statistical estimation of the drift matrix parameter $\mathbb{V}_0$, the quadratic variation of the stochastic component is used.

The OEF is given by the quadratic form [3.2.18], relative to the matrix parameter $\mathbb{V}_0$. *A priori* estimate $\mathbb{V}_T^0$ for the drift matrix $\mathbb{V}_0$ is determined by an optimality ratio (Theorem 3.2.2). The best estimate $\mathbb{V}_T^0$ coincides with *a priori* statistics $\mathbb{V}_T^0 = \mathbb{R}_T^0\,\mathbb{R}_T^{-1}$, which is determined by a connection of covariance matrices.

The minimum OEF determines an estimate of the covariance matrix of the stochastic component [3.2.21].

The Gaussianity of stationary Markov SEs, given by covariance matrices, ensures their representation by the solution of the vector SDEs.

3.2.1. *Vector difference SDEs and stationarity conditions*

DEFINITION 3.2.1.– The MSEs $\alpha_t := (\alpha_t^k\,,\ 1 \leq k \leq M)$, $t \geq 0$, $M > 1$, are given by the solution of the vector SDEs (see section 1.1.4):

$$\Delta\alpha_{t+1} = -\mathbb{V}_0\alpha_t + \Delta\mu_{t+1}, \quad t \geq 0. \qquad [3.2.1]$$

The drift component is determined by the matrix

$$\mathbb{V}_0 := [V_{km}^0;\ 1 \leq k, m \leq M], \qquad [3.2.2]$$

which satisfies the additional stationarity condition

The eigenvalues of the matrix $\mathbb{Q} := \mathbb{I} - \mathbb{V}_0$
are placed strictly inside the unit circle. [3.2.3]

The stochastic component is determined by a square integrable vector stationary martingale $\mu_t = (\mu_t^k,\ 1 \leq k \leq M)$, $t \in \mathcal{N}_0$, which is characterized by the first two moments

$$E\Delta\mu_{t+1} = 0, \quad E[\Delta\mu_{t+1}\Delta\mu_{t+1}{}'] = \mathbb{B}, \quad t \geq 0. \qquad [3.2.4]$$

where $(\cdot)'$ means the transposition of a matrix (or vector) $(\cdot)$.

The characterization of the wide-sense stationarity of the MSEs α_t, $t \geq 0$, gives the following theorem.

THEOREM 3.2.1.– (Stationarity theorem). *A MSE given by the solution of the SDE* [3.2.1], *is a wide-sense stationary process:*

$$E[\alpha_{t+s}\alpha_t{}'] = \mathbb{R}(s) = \mathbb{Q}^s\mathbb{R} \quad \forall t \in \mathcal{N}_0, \qquad [3.2.5]$$

under the following conditions:

A: The positively defined covariance matrix $\mathbb{R}$ *is given by the solution of the matrix equation of stationarity:*

$$\mathbb{B} = \mathbb{V}_0\mathbb{R} + \mathbb{R}\mathbb{V}_0{}' - \mathbb{V}_0\mathbb{R}\mathbb{V}_0{}'. \qquad [3.2.6]$$

B: The initial value α_0*, determined by the first two moments*

$$E\alpha_0 = 0, \quad E[\alpha_0\alpha_0{}'] = \mathbb{R},$$

is a random variable, independent of the martingale differences $\Delta\mu_t$, $t \geq 1$.

Proof of Theorem 3.2.1.– By analogy with Theorem 3.1.2 (see the previous section), the stationary SEs [3.2.1]–[3.2.4] are completely determined by covariance matrices (Koroliouk 2016b):

$$E[\alpha_t\alpha_t'] = \mathbb{R}, \quad E[\Delta\alpha_{t+1}\Delta\alpha_{t+1}'] = \mathbb{R}^{\Delta},$$
$$E[\Delta\alpha_{t+1}\alpha_t'] = \mathbb{R}^{0}, \quad E[\alpha_t\Delta\alpha_{t+1}'] = \mathbb{R}^{0\prime}. \qquad [3.2.7]$$

Using the relationship between covariance matrices $\mathbb{R}^0$ and $\mathbb{R}$, generated by SDE [3.2.1],

$$\mathbb{R}^0 = -\mathbb{V}\mathbb{R}, \quad \mathbb{R}^{0\prime} = -\mathbb{R}\mathbb{V}'. \qquad [3.2.8]$$

However, there is a relation

$$\mathbb{R}^{\Delta} = -\left[\mathbb{R}^{0\prime} + \mathbb{R}^0\right]. \qquad [3.2.9]$$

Indeed, using the definition $\Delta\alpha_{t+1} := \alpha_{t+1} - \alpha_t$, we have the following transformation:

$$\mathbb{R}^{\Delta} = E[\alpha_{t+1}\,\Delta\alpha_{t+1}'] - \mathbb{R}^{0\prime}. \qquad [3.2.10]$$

Using the stationarity of SEs, the first term becomes

$$E[\alpha_{t+1}\,\Delta\alpha_{t+1}'] = -E[\Delta\alpha_{t+1}\,\alpha_t'] = -\mathbb{R}^0. \qquad [3.2.11]$$

So, the covariance matrix $\mathbb{R}^{\Delta}$ has the following representation:

$$\mathbb{R}^{\Delta} = \mathbb{V}\mathbb{R} + \mathbb{R}\mathbb{V}'. \qquad [3.2.12]$$

Finally, we compute the covariant matrix of the stochastic component:

$$\mathbb{B} = E[\Delta\mu_{t+1}\,\Delta\mu_{t+1}'] = E[(\Delta\alpha_{t+1} + \mathbb{V}\alpha_t)(\Delta\alpha_{t+1} + \mathbb{V}\alpha_t)']$$
$$= \mathbb{R}^{\Delta} + \mathbb{R}^0\mathbb{V}' + \mathbb{V}\mathbb{R}^{0\prime} + \mathbb{V}\mathbb{R}\mathbb{V}'. \qquad [3.2.13]$$

Using the representation [3.2.8] of covariance matrices $\mathbb{R}^0$ and $\mathbb{R}^{0\prime}$, we obtain the stationarity condition [3.2.6]. Theorem 3.2.1 is proved. □

REMARK 3.2.1.– The independence of the initial value α_0 from the stochastic component $\Delta\mu_{t+1}$, $t \geq 1$, is used when applying the stationarity condition to prove the relation [3.2.11].

REMARK 3.2.2.– The covariance matrix of the stochastic component [3.2.4] also has representation through the covariances of SE [3.2.7]:

$$\mathbb{B} = \mathbb{R}^{\Delta} - \mathbb{R}^{0}\mathbb{R}^{-1}\mathbb{R}^{0\,\prime}. \qquad [3.2.14]$$

The existence of the inverse matrix $\mathbb{R}^{-1}$ is ensured by the regularity of stationary process α_t, $t \geq 0$ (Bulinski and Shiryaev 2003, Ch. 7, section 15).

3.2.2. *Optimal estimating function*

Consider the following covariance statistics:

$$\mathbb{R}_T := \tfrac{1}{T}\textstyle\sum_{t=1}^{T} \alpha_t \alpha_t{}', \quad \mathbb{R}_T^0 := \tfrac{1}{T}\textstyle\sum_{t=1}^{T} \Delta\alpha_{t+1}\alpha_t'$$
$$\mathbb{R}_T^{0\,\prime} := \tfrac{1}{T}\textstyle\sum_{t=1}^{T} \alpha_t \Delta\alpha_{t+1}', \quad \mathbb{R}_T^{\Delta} := \tfrac{1}{T}\textstyle\sum_{t=1}^{T} \Delta\alpha_{t+1}\Delta\alpha_{t+1}'. \qquad [3.2.15]$$

Following the approach (Heyde 1997) used in the statistical analysis for stationary SEs [3.2.1]–[3.2.4], let us consider the martingale:

$$\mathbb{M}_T(\mathbb{V}) := \sum_{t=0}^{T-1} (\Delta\alpha_{t+1} + \mathbb{V}\alpha_t), \qquad [3.2.16]$$

defined by the stochastic component of equation [3.2.1].

DEFINITION 3.2.2.– The *OEF* is given by the quadratic variation of the martingale: [3.2.16]:

$$[\mathbb{M}(\mathbb{V})]_T := \frac{1}{T}\sum_{t=1}^{T} (\Delta\alpha_{t+1} + \mathbb{V}\alpha_t)(\Delta\alpha_{t+1} + \mathbb{V}\alpha_t)'. \qquad [3.2.17]$$

Then the OEF [3.2.17] has the representation

$$[\mathbb{M}(\mathbb{V})]_T = \mathbb{V}\mathbb{R}_T\mathbb{V}' + \mathbb{V}\mathbb{R}_T^{0\,\prime} + \mathbb{R}_T^0\mathbb{V}' + \mathbb{R}_T^{\Delta}. \qquad [3.2.18]$$

The OEF [3.2.18] admits a representation with a positive-definite component, which is used to construct an optimal statistical estimate of the drift matrix.

PROPOSITION 3.2.1.– The OEF for the true value of the predictable matrix $\mathbb{V}_0$ is determined by the relation

$$[\mathbb{M}(\mathbb{V})]_T = \mathcal{E}_T^0(\mathbb{V}, \mathbb{V}_T^0) + \mathcal{B}_T, \qquad [3.2.19]$$

where the first term on the right-hand side is a positively defined matrix

$$\mathcal{E}_T^0(\mathbb{V}, \mathbb{V}_T^0) = (\mathbb{V} - \mathbb{V}_T^0)\,\mathbb{R}_T\,(\mathbb{V} - \mathbb{V}_T^0)', \qquad [3.2.20]$$

and the residual term has the form

$$\mathcal{B}_T = \mathbb{R}_T^\Delta - \mathbb{V}_T^0 \mathbb{R}_T \mathbb{V}_T^{0\prime}. \qquad [3.2.21]$$

The *a priori* statistic $\mathbb{V}_T^0$ is determined by the relation of two covariance statistics:

$$\mathbb{V}_T^0 = -\mathbb{R}_T^0 \mathbb{R}_T^{-1} \quad \text{or equivalently} \quad \mathbb{R}_T^0 = -\mathbb{V}_T^0 \mathbb{R}_T. \qquad [3.2.22]$$

Proof of Proposition 3.2.1.– The OEF [3.2.18], taking into account relation [3.2.22], turns into the standard form

$$\mathcal{E}_T^0(\mathbb{V}, \mathbb{V}_0) = \mathbb{V}\mathbb{R}_T\mathbb{V}' - \mathbb{V}\mathbb{R}_T\mathbb{V}_T^{0\prime} - \mathbb{V}_T^0\mathbb{R}_T\mathbb{V}' + \mathbb{R}_T^\Delta. \qquad [3.2.23]$$

Finally, we have the representation

$$\mathcal{E}_T^0(\mathbb{V}, \mathbb{V}_0) = (\mathbb{V} - \mathbb{V}_T^0)\,\mathbb{R}_T(\mathbb{V} - \mathbb{V}_T^0)' + \mathbb{R}_T^\Delta - \mathbb{V}_T^0\mathbb{R}_T\mathbb{V}_T^{0\prime}, \qquad [3.2.24]$$

in which the first component is determined by a positively defined function [3.2.20], and the residual term [3.2.21] is determined by the covariance statistics. □

THEOREM 3.2.2.– *The a priori estimate* $\mathbb{V}_T^0$ [3.2.22] *for the drift matrix* $\mathbb{V}_0$ *is determined by the optimality ratio*

$$\min_{\mathbb{V}}[\mathbb{M}(\mathbb{V})]_T = \mathcal{E}_T^0(\mathbb{V}_0, \mathbb{V}_0) + \mathcal{B}_T = \mathcal{B}_T. \qquad [3.2.25]$$

The minimum optimality ratio [3.2.25] *is achieved in the value of residual term* [3.2.21].

3.2.3. *Stationary Gaussian Markov SEs*

In sections 3.2.1 and 3.2.2, the wide-sense stationary SEs are determined by SDEs.

It is known (Nevelson and Has'minskii 1976) that the solution of the SDE [3.2.1] specifies a Markov SE.

It turns out that stationary Gaussian Markov SEs, given by the covariance matrices,

$$E[\alpha_t \alpha_t'] = \mathbb{R}, \quad E[\Delta\alpha_{t+1}\Delta\alpha_{t+1}'] = \mathbb{R}^{\Delta},$$
$$E[\Delta\alpha_{t+1}\alpha_t'] = \mathbb{R}^0, \quad E[\alpha_t \Delta\alpha_{t+1}'] = \mathbb{R}^{0\prime}, \qquad [3.2.26]$$

are determined by the solution of the SDEs.

Without limiting generality, we can assume that

$$E\alpha_t = 0.$$

THEOREM 3.2.3.– (Equivalence theorem). *The two-component Gaussian Markov stationary SE* $(\alpha_t, \Delta\alpha_{t+1})$, $t \geq 0$, *with covariance matrices* [3.2.26], *are given by the solution of the following SDE:*

$$\Delta\alpha_{t+1} = -\mathbb{V}\alpha_t + \Delta W_{t+1}, \quad t \in \mathcal{N}_0. \qquad [3.2.27]$$

The sequence of independent Gaussian martingale differences (ΔW_{t+1}), $t \geq 0$, *is determined by the first two moments:*

$$E\Delta W_{t+1} = 0, \quad E[\Delta W_{t+1}\Delta W_{t+1}'] = \mathbb{B},$$
$$\mathbb{B} = \mathbb{V}\mathbb{R} + \mathbb{R}\mathbb{V}' - \mathbb{V}\mathbb{R}\mathbb{V}'. \qquad [3.2.28]$$

Proof of Theorem 3.2.3.– Using the normal correlation theorem (Liptser and Shiryaev 2001, Theorem 13.1), let us calculate the conditional expectation:

$$E[\Delta\alpha_t \mid \alpha_t] = \mathbb{R}^0\mathbb{R}^{-1}\alpha_t. \qquad [3.2.29]$$

Consider the matrix $\mathbb{V}$:

$$\mathbb{V} := -\mathbb{R}^0\mathbb{R}^{-1}. \qquad [3.2.30]$$

Then the following relations take place:

$$\mathbb{R}^0 = -\mathbb{V}\mathbb{R}, \quad \mathbb{R}^{0\prime} = -\mathbb{R}\mathbb{V}', \qquad [3.2.31]$$

and

$$\mathbb{R}^{\Delta} = \mathbb{V}\mathbb{R} + \mathbb{R}\mathbb{V}'. \qquad [3.2.32]$$

Consider the martingale differences

$$\Delta W_{t+1} := \Delta\alpha_{t+1} + \mathbb{V}\alpha_t, \quad t \geq 0. \qquad [3.2.33]$$

Obviously, according to [3.2.29],

$$E\Delta W_{t+1} = EE[\Delta W_{t+1} \mid \alpha_t] = 0, \quad t \geq 0. \qquad [3.2.34]$$

Now we calculate the covariance of the stationary Gaussian Markovian martingale difference [3.2.33]:

$$\begin{aligned}\mathbb{B} := E[\Delta W_{t+1}\, \Delta W_{t+1}{}'] &= E[(\Delta\alpha_{t+1} + \mathbb{V}\alpha_t)(\Delta\alpha_{t+1} + \mathbb{V}\alpha_t)'] \\ &= \mathbb{R}^{\Delta} + \mathbb{R}^0\mathbb{V}' + \mathbb{V}\mathbb{R}^{0\prime} + \mathbb{V}\mathbb{R}\mathbb{V}'. \qquad [3.2.35]\end{aligned}$$

Using relations [3.2.31], we find the covariance of the stochastic component

$$\mathbb{B} := E[\Delta W_{t+1}\, \Delta W_{t+1}{}'] = \mathbb{R}^{\Delta} - \mathbb{V}\mathbb{R}\mathbb{V}'. \qquad [3.2.36]$$

Taking into account [3.2.32], the covariance of the stochastic component has the representation:

$$\mathbb{B} = \mathbb{V}\mathbb{R} + \mathbb{R}\mathbb{V}' - \mathbb{V}\mathbb{R}\mathbb{V}'. \qquad [3.2.37]$$

Representation [3.2.37] coincides with the stationarity condition [3.2.6] of the stochastic component of the SDE [3.2.1] for a wide-sense stationary SE.

Now it remains, using the Markov property of SEs, to make sure that the sequence of stochastic components [3.2.33] is given by independent Gaussian random variables with dispersion [3.2.37], according to Lemma 13.4 of Liptser and Shiryaev (2001, Ch. 13, section 1), that is,

$$E[\Delta W_{t+1}\Delta W_{s+1}{}'] = 0 \quad \text{by} \quad t \neq s. \qquad [3.2.38]$$

Suppose for determination $s < t$. Using the Markov property of SEs, as well as [3.2.34], we have

$$\begin{aligned}E[\Delta W_{t+1} \mid \alpha_t, \alpha_s] &= E[\Delta\alpha_{t+1} + \mathbb{V}\alpha_t \mid \alpha_t, \alpha_s] \\ &= E[\Delta\alpha_{t+1} + \mathbb{V}\alpha_t \mid \alpha_t] = 0. \qquad [3.2.39]\end{aligned}$$

Consequently, for $s < t$,

$$\begin{aligned}E[\Delta W_{t+1}\, \Delta W'_{s+1}] &= E[E[\Delta W_{t+1} \mid \alpha_t] \cdot \Delta W_{s+1} \mid \alpha_s] \\ &= E[\Delta W_{t+1} \mid \alpha_t] \cdot E[\Delta W'_{s+1} \mid \alpha_s] = 0. \qquad [3.2.40]\end{aligned}$$

Then, formula [3.2.38] is proved.

Finally, the stationary Gaussian Markov SEs, characterized by the covariance matrix [3.2.36], are given by the solution of the SDE [3.2.27]. The stochastic component is determined by the independent Gaussian martingale differences with the parameters

$$E\Delta W_{t+1} = 0, \quad E[\Delta W_{t+1}\Delta W_{t+1}'] = \mathbb{B}. \qquad [3.2.41]$$

Theorem 3.2.3 is proved. □

3.3. Estimates of continuous process parameters

3.3.1. *Diffusion-type processes*

It is known (Liptser and Shiryaev 2001) that the diffusion-type processes are given by the stochastic differential

$$\mathrm{d}\xi_t = \alpha_t(\xi)\,\mathrm{d}t + \mathrm{d}W_t. \qquad [3.3.1]$$

The drift component satisfies the following conditions:

$$\begin{aligned} &P\left(\int_0^T \alpha_t^2(\xi)\,\mathrm{d}t < \infty\right) = 1, \quad T < \infty, \\ &P\left(\int_0^\infty \alpha_t^2(\xi)\,\mathrm{d}t = \infty\right) = 1, \end{aligned} \qquad [3.3.2]$$

which ensures absolute continuity of measures $\mu_\xi(B) := P\{\omega : \xi \in B\}$ and $\mu_W(B) := P\{\omega : W \in B\}$ for all $B \in \mathcal{B}_T$ in the space of measurable continuous functions $(\mathcal{C}_T, \mathcal{B}_T)$ on the interval $0 \le t \le T$.

The derivative of Radon–Nikodym specifies the density of the measure

$$\zeta_T(\xi) := \frac{\mathrm{d}\mu_\xi}{\mathrm{d}\mu_W}(\xi, T), \qquad [3.3.3]$$

which, for the diffusion-type processes [3.3.1], has the following representation.

THEOREM 3.3.1.– (Liptser and Shiryaev 2001, §7.2, Theorem 7.6) *The measure density* [3.3.3] *of the diffusion-type process* [3.3.1] *with additional conditions* [3.3.2] *is given by the exponential martingale*

$$\frac{\mathrm{d}\mu_\xi}{\mathrm{d}\mu_W}(\xi, T) = \exp\left[\int_0^T \alpha_t(\xi)\,\mathrm{d}\xi_t - \frac{1}{2}\int_0^T \alpha_t^2(\xi)\,\mathrm{d}t\right]. \qquad [3.3.4]$$

In particular, the exponential martingale [3.3.3] is determined by the solution of the stochastic Doléans–Dade equation (Borovskikh and Korolyuk 1997):

$$d\zeta_T(\xi) = \zeta_T(\xi)\alpha_T(\xi)\,d\xi_T, \quad \zeta_0(\xi) = 1, \tag{3.3.5}$$

or in equivalent form

$$\zeta_T(\xi) - 1 = \int_0^T \zeta_t(\xi)\alpha_t(\xi)\,d\xi_t. \tag{3.3.6}$$

The relationship of the density [3.3.4] with the stochastic Doléans–Dade equation [3.3.5] is explained using the Ito formula for the exponential function $\varphi(\xi) = \exp[\eta_T(\xi) - \frac{1}{2}\langle\eta(\xi)\rangle_T]$, with $\eta_T(\xi) := \int_0^T \alpha_t(\xi)\,dt$, $\langle\eta(\xi)\rangle_T := \int_0^T \alpha_t^2(\xi)\,dt$, namely (Borovskikh and Korolyuk 1997):

$$d\varphi(\xi_T) = \varphi'(\xi_T)\left[d\eta_T(\xi) - \frac{1}{2}d\langle\eta(\xi)\rangle_T\right] + \frac{1}{2}\varphi''(\xi_T)\,d\langle\eta(\xi)\rangle_T. \tag{3.3.7}$$

Taking into account the equality $\varphi(\xi_T) = \varphi'(\xi_T) = \varphi''(\xi_T)$ for exponential function $\varphi(\xi)$, we obtain the stochastic Doléans–Dade differential equation for the exponential martingale [3.3.3].

3.3.2. *Estimation of a continuous parameter*

According to the results in section 3.2, the limit diffusion process for normalized discrete Markov processes (DMPs) is the Ornstein–Uhlenbeck process with a linear predictable component

$$d\alpha_t = -V_0\alpha_t\,dt + \sigma\,dW_t, \quad 0 \le t \le T, \tag{3.3.8}$$

Without limiting of generality, let us put $\sigma = 1$.

The *method of maximum likelihood* of the parameter V_0 estimation for a diffusion process with stochastic differential [3.3.8] ($\sigma = 1$) is realized for the logarithm of the density [3.3.4]:

$$L(V,T) := \ln\,\zeta_T(\xi) = -V\int_0^T \alpha_t\,dt - \frac{V^2}{2}\int_0^T \alpha_t^2\,dt. \tag{3.3.9}$$

So, the equation for the maximum likelihood estimation is

$$\max_{0\le V\le 2} \frac{\partial L(V,T)}{\partial V} = -\int_0^T \alpha_t\,dt - V_T\int_0^T \alpha_t^2\,dt = 0 \tag{3.3.10}$$

and the maximum likelihood estimation is

$$V_T = -\int_0^T \alpha_t \, \mathrm{d}\alpha_t \bigg/ \int_0^T \alpha_t^2 \, \mathrm{d}t. \qquad [3.3.11]$$

The least-square method of the parameter V_0 estimation for a diffusion process with stochastic differential [3.3.1] ($\sigma = 1$) is realized using the equality

$$\int_0^T \alpha_t \, \mathrm{d}\alpha_t = -V_0 \int_0^T \alpha_t^2 \, \mathrm{d}t + \int_0^T \alpha_t \, \mathrm{d}W_t. \qquad [3.3.12]$$

Hence, we have a relationship

$$\min_{0 \le V_T \le 2} E[V_0 - V_T]^2 \int_0^T \alpha_t^2 \, \mathrm{d}t = E[V_T^0 - V_T]^2 \int_0^T \alpha_t^2 \, \mathrm{d}t. \qquad [3.3.13]$$

The estimation of the least-squares method has a representation

$$V_T^0 = -\frac{\int_0^T \alpha_t \, \mathrm{d}\alpha_t}{\int_0^T \alpha_t^2 \, \mathrm{d}t}. \qquad [3.3.14]$$

Conclusion 3.3.1. Estimates by the maximum likelihood method and the least-squares method coincide: $V_T = V_T^0$.

Conclusion 3.3.2. Estimation by the least-squares method and hence estimation by the method of maximum likelihood of the parameter V_0 are *strictly consistent*:

$$P1 \lim_{T \to \infty} V_T^0 = V_0. \qquad [3.3.15]$$

3.4. Classification of EPs

In section 1.2, the EPs were considered. It was shown that *the fundamental principle of the dynamics of EPs* —"stimulation deterrence"—determines the dynamics of the increments at each stage, by the values of equilibriums $\rho_\pm$ in the range (0,1) with a coefficient of proportionality $V_0 > 0$.

However, the real SE and their EPs can be determined in the admissibility domains of drift parameters, which significantly change the behavior of SEs and their EPs. In section 3.4, we considered the classification of EPs, arises from the consideration of the parameter V_0 and equilibriums $\rho_\pm$, including the binary equilibrium $\rho = \rho_+ - \rho_-$.

We recall the basic assumptions (Propositions 1.2.1 and 1.2.2), in which the dynamics of EPs is determined.

3.4.1. *Basic assumption*

The frequency *evolutionary process* (EP) is given by the conditional expectation (see section 1.2.1):

$$P_\pm(k+1) := E[S_N^\pm(k+1) \mid S_N^\pm(k) = P_\pm(k)], \quad k \geq 0, \qquad [3.4.1]$$

Basic assumption 1. Frequency EPs [3.4.1] are determined by the solutions of the difference evolutionary equations (DEEs) with the regression function of increments (RFI) $V_0(p_+, p_-)$:

$$\begin{aligned} &\Delta P_\pm(k+1) = -V_0(P_+(k),\, P_-(k)), \\ &V_0(p_+, p_-) = \pm V p_+ p_- (p_\pm - \rho_\pm), \quad \rho_+ + \rho_- = 1. \end{aligned} \qquad [3.4.2]$$

Recall (see section 1.2) that the binary SEs are given by averaged sums of sample values

$$S_N(k) := \frac{1}{N}\sum_{n=1}^{N} \delta_n(k), \quad k \geq 0, \qquad [3.4.3]$$

independent and equally distributed, by n, random variables $\delta_n(k), 1 \leq n \leq N$, $k \geq 0$, which take binary values ± 1 with probabilities $P_\pm(k)$, $k \geq 0$.

So, the binary SE has the representation

$$S_N(k) = S_N^+(k) - S_N^-(k), \quad k \geq 0. \qquad [3.4.4]$$

At the same time, there is the obvious identity

$$S_N^+(k) + S_N^-(k) \equiv 1 \quad \forall k \geq 0, \qquad [3.4.5]$$

and the following connection between the binary and frequency SEs:

$$S_N^\pm(k) = \frac{1}{2}\left[1 \pm S_N(k)\right], \quad k \geq 0. \qquad [3.4.6]$$

Now the conditional expectations [3.4.1] are connected with the representation of the binary EPs by the frequency difference:

$$C(k) = P_+(k) - P_-(k), \quad k \geq 0. \qquad [3.4.7]$$

Basic assumption 2. The binary EPs [3.4.6] are determined by the solution of the following difference evolution equation (DEE):

$$\Delta C(k+1) = -V(C(k)), \quad k \geq 0. \qquad [3.4.8]$$

The RFI is

$$V(c) = V(1-c^2)(c-\rho). \qquad [3.4.9]$$

So, we can assert the following.

– According to Basic assumption 1, the frequency EPs [3.4.1] are characterized by the parameter V and by the equilibria $\rho_\pm$ $(\rho_+ + \rho_- = 1)$.

– The binary EPs [3.4.7] (basic assumption 2) are characterized by parameter V and by equilibriums difference $\rho = \rho_+ - \rho_-$.

Taking into account the balance condition $\rho_+ + \rho_- = 1$, the frequency equilibria can be represented through the equilibrium of binary EPs:

$$\rho_\pm = \frac{1}{2}\left[1 \pm \rho\right]. \qquad [3.4.10]$$

3.4.2. *Classification of EPs*

The behavior of alternative probabilities, determined by the SE dynamics, essentially depends on the values of the corresponding parameters.

The classification of SE models, based on the probabilities of alternatives $P_\pm(k)$, $k \geq 0$, is presented in the following proposition.

We give a classification of frequency EPs taking into account the values of parameter V, as well as the equilibria $\rho_\pm$ $(\rho_+ + \rho_- = 1)$.

PROPOSITION 3.4.1.– Equilibrium values of alternatives $\rho_\pm$ generate the classification of SE models, determined by the DEE solutions [3.4.2]:

$$\begin{array}{llll} \mathbb{M}A & \text{attractive}: & V > 0, & 0 < \rho_\pm < 1, \\ \mathbb{M}R & \text{repulsive}: & V < 0, & 0 < \rho_\pm < 1, \\ \mathbb{M}+ & \text{dominant}+: & V \neq 0, & V\rho_- \leq 0 \leq V\rho_+, \\ \mathbb{M}- & \text{dominant}-: & V \neq 0, & V\rho_+ \leq 0 \leq V\rho_-. \end{array}$$

At the same time, it is possible to give an equivalent classification of binary EPs, taking into account the values of the binary equilibria $\rho = \rho_+ - \rho_-$.

PROPOSITION 3.4.2.– Equilibrium values of alternatives $\rho = \rho_+ - \rho_-$ generate the following classification of models which in turn generate the following classification of the DEE solutions [3.4.8]:

$$\begin{array}{llll} \mathbb{M}A & \text{attractive}: & |\rho| \leq 1, & V > 0; \\ \mathbb{M}R & \text{repulsive}: & |\rho| \leq 1, & V < 0; \\ \mathbb{M}+ & \text{dominant}+: & |\rho| \geq 1, & V \cdot \rho > 0; \\ \mathbb{M}- & \text{dominant}-: & |\rho| \geq 1, & V \cdot \rho < 0. \end{array}$$

REMARK 3.4.1.– The classification of frequency EPs, given by the probabilities of alternatives $P_\pm(t)$, $t \geq 0$ (in continuous time) with the regression function (RF) $Q(x) = x(1 - x)(ax + b)$, is contained in Skorokhod *et al.* (2002) and corresponds to the classification given in Proposal 3.4.2.

3.4.3. *Justification of EP models classification*

The basis for SE classification, presented in Propositions 3.4.1 and 3.4.2, is the limit behavior of probabilities of SE alternatives, formulated in the following theorem.

THEOREM 3.4.1.– *Probabilities of alternatives $P_\pm(k)$, $k \geq 0$, determined by the DEE solutions* [3.4.2] *or* [3.4.8], *have the following asymptotic behavior:*

■ *In the model $\mathbb{M}A$ (attractive)*

$$\lim_{k\to\infty} P_\pm(k) = \rho_\pm. \qquad [3.4.11]$$

■ *In the model* $\mathbb{M}R$ *(repulsive)*

$$\lim_{k\to\infty} P_{\pm}(k) = 1, \qquad \textit{or equivalently,} \qquad \lim_{k\to\infty} P_{\mp}(k) = 0, \qquad [3.4.12]$$

under initial conditions

$$P_{\pm}(0) > \rho_{\pm}, \qquad \textit{or equivalently,} \qquad P_{\mp}(0) < \rho_{\mp}. \qquad [3.4.13]$$

■ *In the model* $\mathbb{M}\pm$ *(domination*$\pm$*)*

$$\lim_{k\to\infty} P_{\pm}(k) = 1, \qquad \textit{or equivalently,} \qquad \lim_{k\to\infty} P_{\mp}(k) = 0. \qquad [3.4.14]$$

Of interest is the interpretation, as well as the motivation of different SE models in practical applications, such as population genetics (Svirezhev and Pasekov 1982) and economics and behavior patterns (Bousquet *et al.* 2004, Shiryaev 1999, Vapnik 2000).

For example, let us consider the interpretation of SE classification in terms of behavioral patterns.

Taking into account the RFI [3.4.2], the classification of SE models is determined by the values of drift parameters $V_{\pm}$, which characterize *stimulation and deterrence* of increments of alternatives probabilities (Koroliouk 2015a).

So, in the model $\mathbb{M}A$ (attractive), $V > 0$, $0 < \rho_{\pm} < 1$, the probability of positive alternative $P_{+}(k)$, $k \geq 0$, increases (stimulated) in proportion to the probability of negative alternative with the parameter V_{-}, and they are diminished (deterred) in proportion to the probability of a positive alternative with the parameter V_{+}. Such *a characterization of stimulation and deterrence* conducts to existence of a steady regime, determined by the equilibria $\rho_{\pm}$ of the RFI.

But in the model $\mathbb{M}R$ (repulsive), $V < 0, 0 < \rho_{\pm} < 1$. The characterization of stimulations and deterrences has a reverse effect on the probabilities of alternatives. As a result, the equilibrium of regression increments becomes *repulsive*, that is, the probabilities of alternatives are directed to absorbing states $\rho_{\pm} = 0$ or 1. The repulsive equilibrium state $\rho_{\pm}$ plays the role of limiting threshold. At initial probability of alternatives less than this threshold, the next probabilities are conducted, with diminishing, to the absorbing state $\rho = 0$. And with the initial probability of alternatives greater than the threshold $\rho_{\pm}$, the next probabilities increase up to the absorbing state $\rho = 1$.

Finally, in the model $\mathbb{M}\pm$ (dominant $\pm$), $|\rho_{+} - \rho_{-}| \geq 1$, the equilibrium of RFI $\rho_{\pm}$ is outside the interval $(0, 1)$. Thus, the probabilities of alternatives, going (pulling or pushing) to the equilibria, eventually reach the absorbing states $\rho_{\pm} = 0$ or 1.

In the model $\mathbb{M}+$ (dominant $+$), the probabilities of positive alternatives tend to 1, while the probability of negative alternatives tends to 0.

But in the model $\mathbb{M}-$ (dominant $-$), the probabilities of negative alternatives tend to 1.

3.4.4. *Proof of Theorem 3.4.1*

The main idea is that the probabilities of the alternatives, given by the solutions of the DEEs [3.4.2] or [3.4.8], are bounded and monotone for $k \to \infty$. This ensures the existence of limits. It remains only to make sure that the limit relations for frequency probabilities derived from the DEE [3.4.2] or [3.4.8] provide the assertion of the theorem.

The proof is based on the following DEE (see [3.4.2]):

$$\Delta \widehat{P}_{\pm}(k+1) = -V P_{+}(k) P_{-}(k) \widehat{P}_{\pm}(k), \quad k \geq 0. \qquad [3.4.15]$$

■ The model $\mathbb{M}A$.

Let us first consider the case $\mathbb{M}A+$, that is, $\rho_{+} < P_{+}(0) < 1$. Then, according to [3.4.2],

$$0 < \widehat{P}_{+}(0) < \rho_{-}. \qquad [3.4.16]$$

From the basic DEE [3.4.15] by $V > 0$, there follows the monotonicity of the sequence $\widehat{P}_{+}(k)$:

$$\widehat{P}_{+}(k+1) < \widehat{P}_{+}(k), \quad k \geq 0. \qquad [3.4.17]$$

Let us rewrite the DEE [3.4.15] in the following form:

$$\widehat{P}_+(k+1) = \widehat{P}_+(k)[1 - VP_+(k)P_-(k)], \quad k \geq 0. \qquad [3.4.18]$$

The obvious inequality

$$|VP_+P_-| \leq \frac{1}{2}. \qquad [3.4.19]$$

implies, taking into account [3.4.18]–[3.4.19], the estimate

$$\widehat{P}_+(k+1) \geq \frac{1}{2}\widehat{P}_+(k) \geq 0, \quad k \geq 0. \qquad [3.4.20]$$

The monotonicity of probabilities [3.4.17], together with estimate [3.4.20], ensures the existence of the limits

$$\widehat{P}^*_\pm := \lim_{k\to\infty} \widehat{P}_\pm(k), \quad P^*_\pm := \lim_{k\to\infty} P_\pm(k), \qquad [3.4.21]$$

for which the following equation holds:

$$P^*_+P^*_-\widehat{P}^*_\pm = 0, \qquad [3.4.22]$$

which implies

$$\widehat{P}^*_\pm = 0, \quad \text{that is,} \quad P^*_+ = \rho_+, \quad \text{and also} \quad P^*_- = \rho_-. \qquad [3.4.23]$$

In the case $\mathbb{M}A-$, that is, $0 < P_+(0) < \rho_+$, by virtue of the relation

$$\widehat{P}_+(k) + \widehat{P}_-(k) = 0, \quad k \geq 0, \qquad [3.4.24]$$

there is a dual task for $\widehat{P}_-(k)$: there is an inequality (also see equation [3.4.16]):

$$\rho_- < P_-(0) < 1, \quad \text{that is,} \quad 0 < \widehat{P}_-(0) < \rho_+.$$

All the above considerations are valid for $\widehat{P}_-(0)$ instead of $\widehat{P}_+(0)$. Consequently, in this case, we obtain the following limit values:

$$\widehat{P}^*_- = 0, \quad \text{that is,} \quad P^*_- = \rho_-, \quad \text{and also} \quad P^*_+ = \rho_+, \qquad [3.4.25]$$

which completes the Proof of Theorem 3.4.1 for the model $\mathbb{M}A$.

■ The model $\mathbb{M}R$.

Let us consider first the case $\mathbb{M}R+$, when $\rho_+ < P_+(0) < 1$. According to [3.4.16],

$$0 < \widehat{P}_+(k) < \rho_-. \qquad [3.4.26]$$

By virtue [3.4.15] for $V < 0$, there follows the monotonicity of the sequence $\widehat{P}_+(k)$:

$$\widehat{P}_+(k+1) > \widehat{P}_+(k), \quad k \geq 0. \qquad [3.4.27]$$

Relations [3.4.15] and [3.4.19], taking into account $V < 0$, give the inequality

$$\widehat{P}_+(k+1) \leq \frac{3}{2}\widehat{P}_+(k), \quad k \geq 0. \qquad [3.4.28]$$

which, by induction, gives the uniform estimate

$$\widehat{P}_+(k+1) \leq \frac{3}{2}\rho_-. \qquad [3.4.29]$$

So, the limits

$$\widehat{P}^*_\pm := \lim_{k\to\infty} \widehat{P}_\pm(k), \quad P^*_\pm := \lim_{k\to\infty} P_\pm(k), \qquad [3.4.30]$$

exist for which, by virtue of equation [3.4.22], the limit values are taking place:

$$P^*_+ = 1, \quad \text{and also} \quad P^*_- = 0. \qquad [3.4.31]$$

In the case $\mathbb{M}R-$, that is, $0 < P_+(0) < \rho_+$, because of [3.4.24], there is a dual problem for $P_-(0)$: there is the inequality (also see equation [3.4.21]):

$$\rho_- < P_-(0) < 1, \quad \text{that is,} \quad 0 < \widehat{P}_-(0) < \rho_+. \qquad [3.4.32]$$

So, all the above considerations of the model $\mathbb{M}R+$ remain valid for $\widehat{P}_-(0)$. Consequently, in this case, we obtain the following limit values:

$$P^*_- = 1, \quad \text{and also} \quad P^*_+ = 0, \qquad [3.4.33]$$

which completes the Proof of Theorem 3.4.1 for the model $\mathbb{M}R$.

■ The model $\mathbb{M}\pm$.

Consider first the case $\mathbb{M}A+$, that is, $V > 0$, $\rho_+ \geq 1$. Consequently, the following assessment takes place:

$$-\rho_+ < \widehat{P}_+(k) < \rho_- < 0, \quad k \geq 0. \qquad [3.4.34]$$

The monotone growth of the sequence

$$\widehat{P}_+(k+1) > \widehat{P}_+(k), \quad k \geq 0 \qquad [3.4.35]$$

follows from the basic equation [3.4.15], and also from the inequalities for the fluctuations [3.4.29].

The upper boundness

$$\widehat{P}_+(k+1) \leq \frac{3}{2}\rho_- \qquad [3.4.36]$$

follows from inequalities [3.4.20] and [3.4.29].

So, the limits [3.4.25] exist and the limit equation [3.4.22] takes place.

Note that $\widehat{P}_+^* \neq 0$, because it is strictly negative by [3.4.29]. The inequality $P_+^* \neq 0$ takes place, because of $P_+(k+1) > P_+(k) > 0$. So, $P_-^* = 0$, that is, $\lim_{k\to\infty} P_+(k) = 1$, which completes the proof for the case $\mathbb{M}A+$.

In the case, $\mathbb{M}A-$, that is, $V < 0$, $\rho_+ \geq 1$ (see [3.4.19]), the dual problem appears (see [3.4.19]) and the following inequality takes place:

$$-\rho_- \leq \widehat{P}_-(k) \leq \rho_+. \qquad [3.4.37]$$

So, all the above considerations are for the probability of the negative alternative. Consequently, there implies the following relationship:

$$P_-^* = 1, \quad \text{and also} \quad P_+^* = 0, \qquad [3.4.38]$$

which completes the Proof of Theorem 3.4.1 for the model $\mathbb{M}R-$.

Similarly, Theorem 3.4.1 is proved in the case $\mathbb{M}R-$ and also in the dual case $\mathbb{M}A-$.

Theorem 3.4.1 is proved. □

3.4.5. *Interpretation of EPs*

Let us give some comments on the justification of classification (section 3.4.2).

The behavior of EPs essentially depends on the nature of the RFI at different values of drift parameters.

Without generality limitation, the frequency evolutionary behavior process $P_+(k)$, $k \geq 0$, is considered as the main one.

In Figure 3.1, the RFI $V_+(p)$ is depicted in the model $\mathbb{M}A$ – attracting equilibrium.

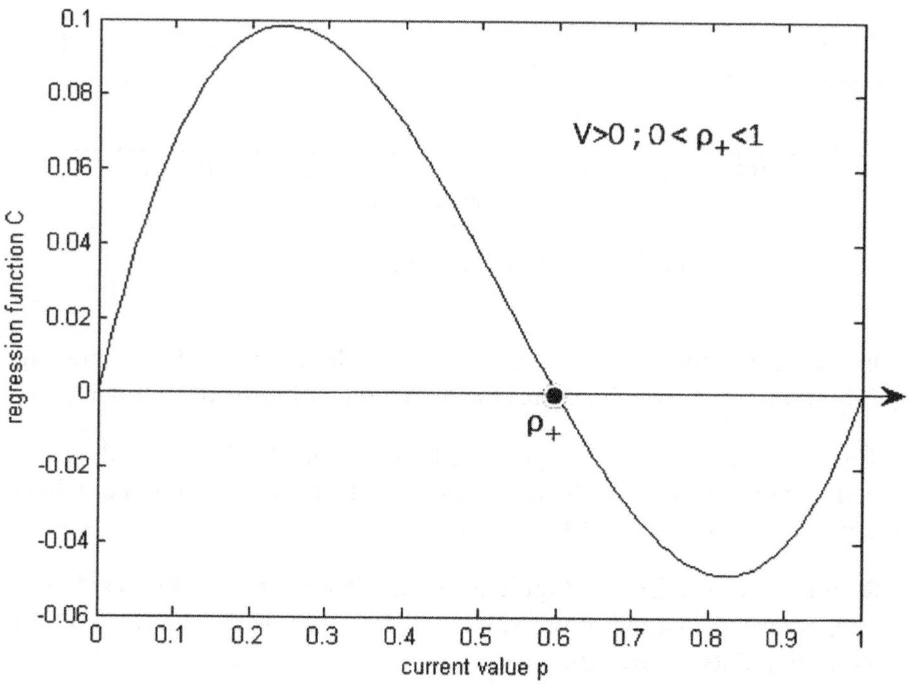

Figure 3.1. *The function $V_+(p)$ in the model $\mathbb{M}A$. For a color version of the figures in this book, see http://www.iste.co.uk/koroliouk/dynamics.zip*

In Figure 3.2, the RFI $V_+(p)$ is depicted in the model $\mathbb{M}R$: – repulsive equilibrium.

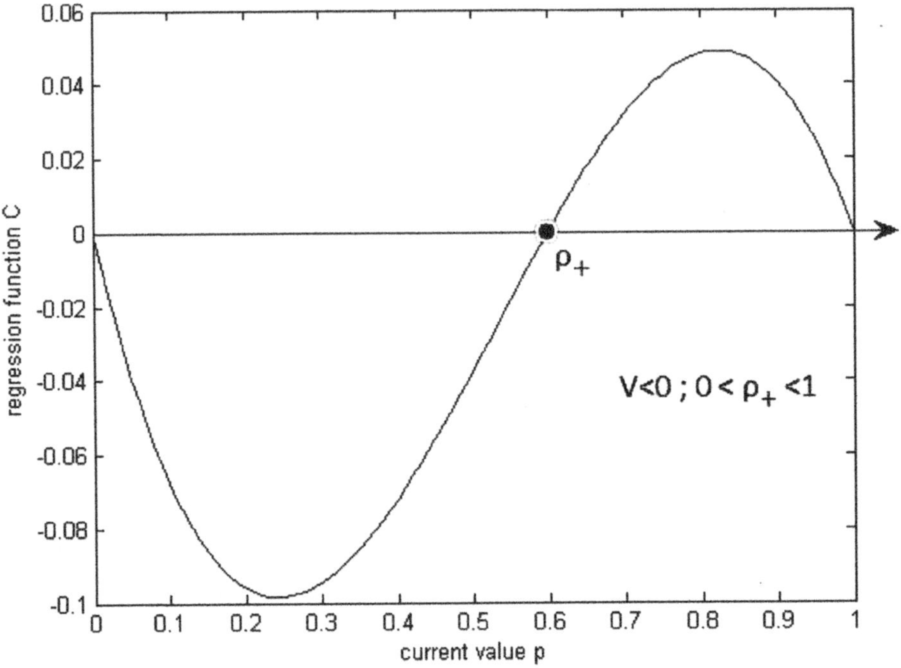

Figure 3.2. *The function $V_+(p)$ in the model $\mathbb{MR}$*

In Figure 3.3, the RFI $V_+(p)$ is depicted in the model $\mathbb{MA}+$ – attracting equilibrium $\rho_+ > 1$, which ensures the domination of positive alternatives.

The classification of EPs, presented in section 3.4.4, is based on the properties of the RFI. Now the illustration of EPs is continued in the light of the behavior of increments $\Delta P_\pm(k)$, as $k \to \infty$.

■ In the model $\mathbb{MA}$ (see Figure 3.5), the alternative probabilities $P_\pm(k)$, $k \geq 0$, converge to equilibrium values $\rho_\pm$ for any initial value $P_\pm(0)$. The equilibrium points $\rho_\pm$ are *attractive*.

Namely, the following limits take place

$$\lim_{k\to\infty} P_\pm(k) = \rho_\pm. \qquad [3.4.39]$$

■ In the model $\mathbb{MR}$ (see Figure 3.6), the asymptotics of alternative probabilities $P_\pm(k)$, as $k \to \infty$, depend on the initial conditions. The equilibrium points $\rho_\pm$ are *repulsive*.

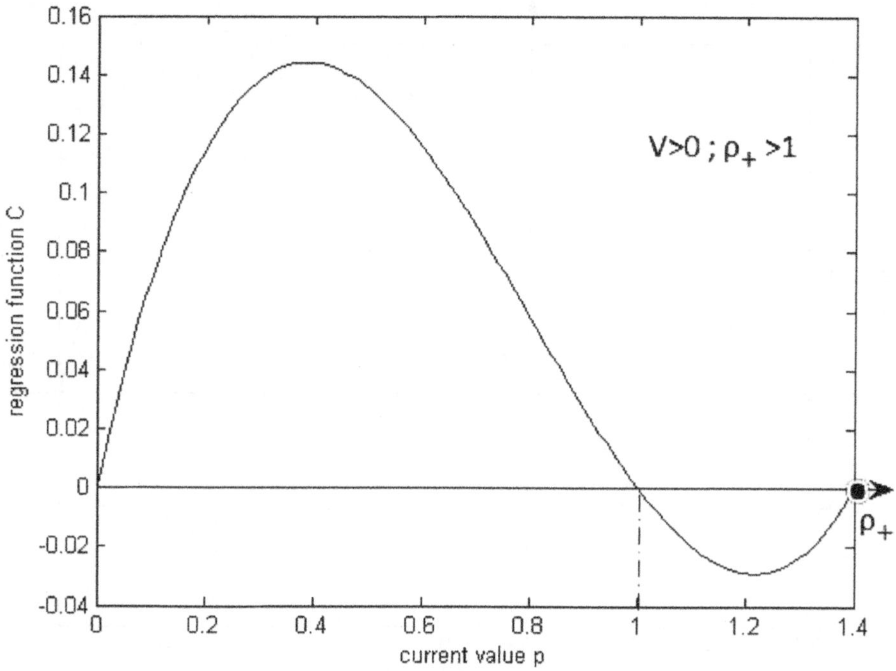

Figure 3.3. *The function $V_+(p)$ in the model* $\mathbb{M}A+$

In particular, the following takes place:

$$\begin{aligned} &\lim_{k\to\infty} P_\pm(k) = 0, \quad \text{if} \quad P_\pm(0) < \rho_\pm, \\ &\lim_{k\to\infty} P_\pm(k) = 1, \quad \text{if} \quad P_\pm(0) > \rho_\pm. \end{aligned} \qquad [3.4.40]$$

■ In the models $\mathbb{M}A+$ and $\mathbb{M}R+$ (dominant $+$), for any initial value $P_\pm(0)$ there is an advantage of the alternative "+": $P_+(k) \to 1$ as $k \to \infty$, while $P_-(k) \to 0$ as $k \to \infty$.

In this case, there are two different situations.

◇ If $V > 0$ (see Figure 3.7), there is an attraction of positive alternative probability to equilibrium $\rho_+ \geq 1$.

More specifically,

$$\lim_{k\to\infty} P_+(k) = 1, \quad \lim_{k\to\infty} P_-(k) = 0. \qquad [3.4.41]$$

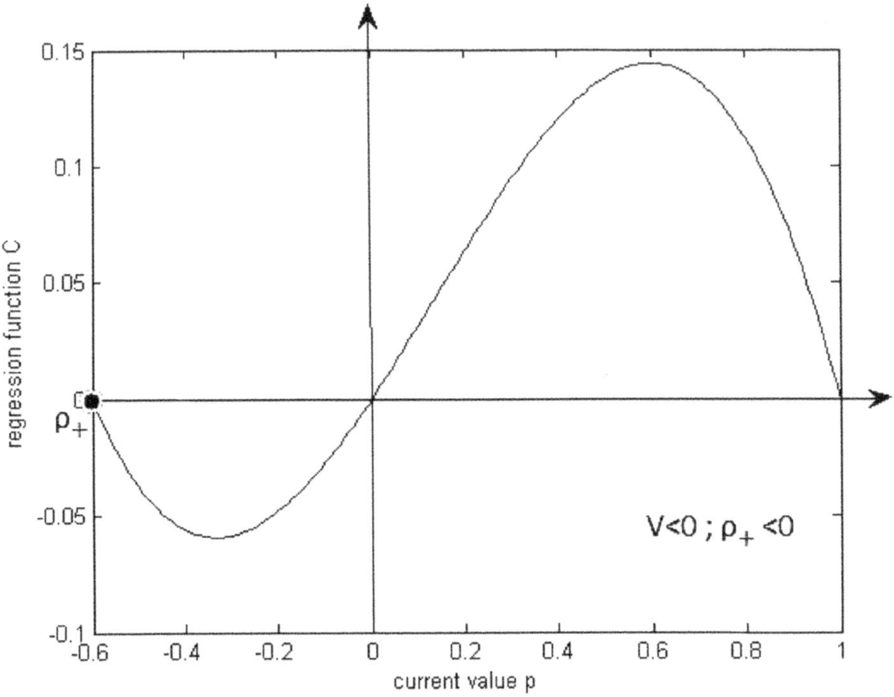

Figure 3.4. *The function $V_+(p)$ in the model $\mathbb{M}R+$*

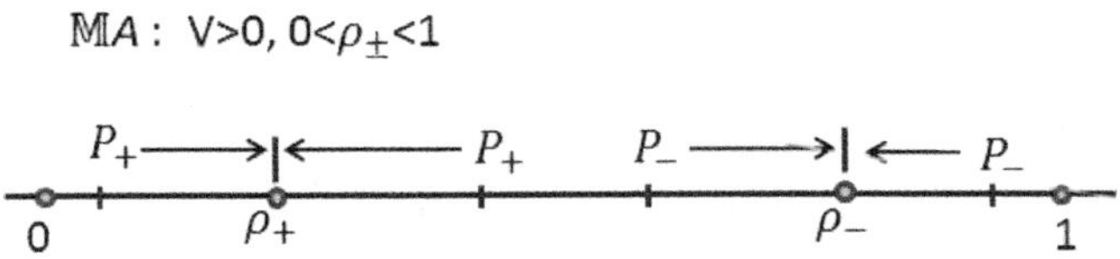

Figure 3.5. *The asymptotic behavior of $P_\pm(k)$ in the model $\mathbb{M}A$ with increasing k*

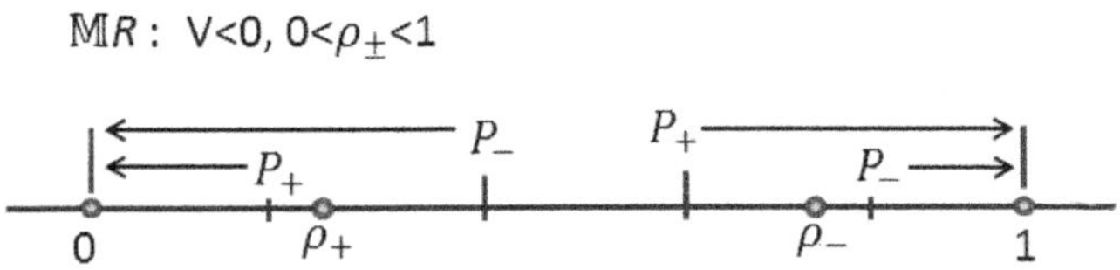

Figure 3.6. *The asymptotic behavior of $P_\pm(k)$ in the model $\mathbb{M}R$ with increasing k*

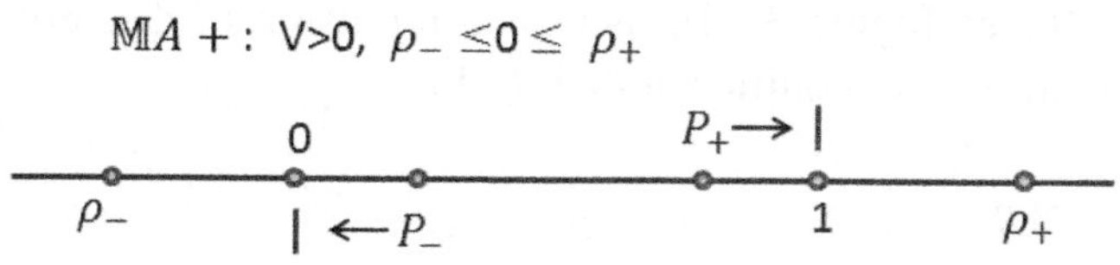

Figure 3.7. *The asymptotic behavior of $P_{\pm}(k)$ in the model $\mathbb{MA}+$ by increasing k*

◇ If $V < 0$ (see Figure 3.8), there is a repulsion of the probability of positive alternatives from equilibrium $\rho_+ \leq 0$.

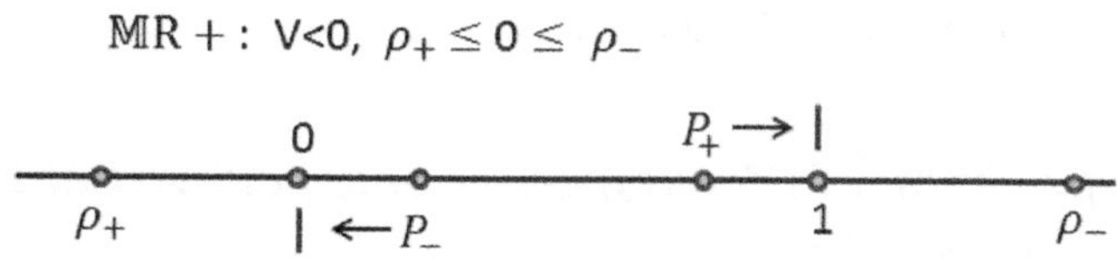

Figure 3.8. *The asymptotic behavior of $P_{\pm}(k)$ in the model $\mathbb{MR}+$ by increasing k*

Again, the convergences [3.4.41] take place.

■ In the models $\mathbb{MA}-$ and $\mathbb{MR}-$ (dominant $-$), for any initial value $P_{\pm}(0)$, the advantage appears for the alternative "$-$": $P_-(k) \to 1$ by $k \to \infty$, while $P_+(k) \to 0$ as $k \to \infty$.

Also, in this case, there are two different situations.

◇ If $V > 0$ (see Figure 3.9), the attraction of the positive alternative probability to equilibrium $\rho_+ \leq 0$ takes place.

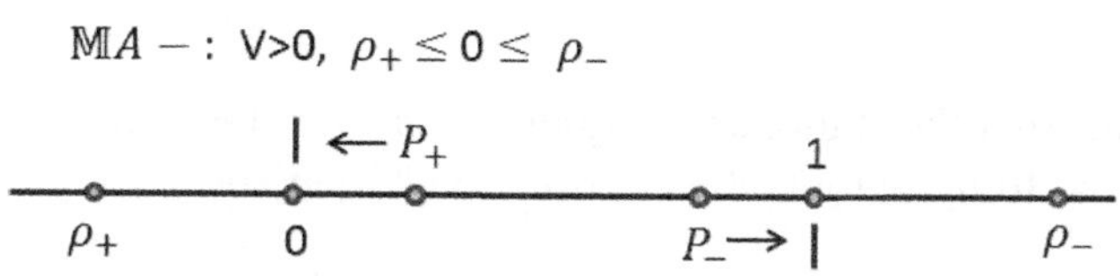

Figure 3.9. *The asymptotic behavior of $P_{\pm}(k)$ in the model $\mathbb{MA}-$ by increasing k*

The following limits take place:

$$\lim_{k\to\infty} P_+(k) = 0, \quad \lim_{k\to\infty} P_-(k) = 1. \qquad [3.4.42]$$

◇ If $V < 0$ (see Figure 3.10), there is a repulsion of the probability of a positive alternative from equilibrium $\rho_+ \geq 1$

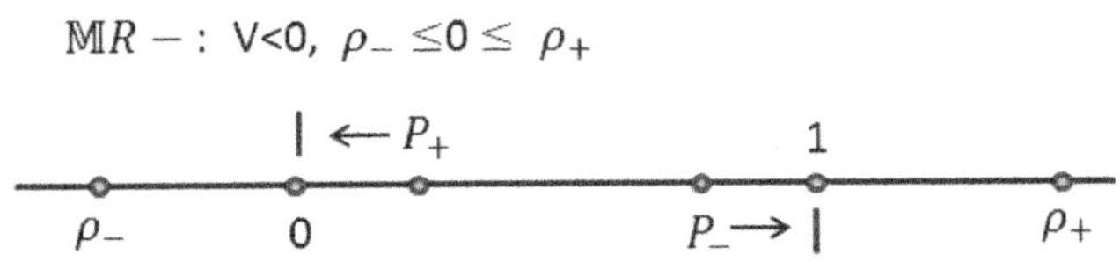

Figure 3.10. *The asymptotic behavior of $P_\pm(k)$ in the model $\mathbb{M}R-$ by increasing k*

Again, the relation [3.4.42] takes place.

3.4.6. *Interpretation of EPs in models of collective behavior*

The classification of EPs can now be used to interpret the collective behavior of N actors adopting the alternative decisions ± 1. Consequently, the probabilities of frequency EPs has the following statistical interpretation (on the kth step):

$$P_\pm(k) = \frac{N_\pm(k)}{N}, \quad N_+(k) + N_-(k) = N, \quad k \geq 0. \qquad [3.4.43]$$

■ In the model $\mathbb{M}A$, there is an attracting point of equilibrium

$$\rho_\pm = \frac{\nu_\pm}{N}, \quad \nu_+ + \nu_- = N, \quad k \geq 0. \qquad [3.4.44]$$

The alternative probabilities $P_\pm(k)$ converge, as $k \to \infty$, to equilibrium values $\rho_\pm$.

This means that the stationary frequencies $\rho_\pm$ for the choice of alternatives that correspond to $\nu_\pm$ actors who choose the alternatives ± 1.

■ In the model $\mathbb{M}R$, the repulsive equilibrium points of alternatives $\rho_\pm$ mean that at the initial value $N_+(0) < \nu_+$, the number of subjects accepting the positive alternative is reduced to zero. And when $N_+(0) > \nu_+$, the number of subjects accepting the positive alternatives becomes dominant: $N_+(k) \to N$ by $k \to \infty$. Accordingly, there are two absorbing points of equilibrium: 0 and 1.

This law seems to deserve the greatest attention of specialists. The next interpretation of this law can be given as an example in the education sector.

Each class has a certain average intelligence ρ, $0 < \rho < 1$. The result of a long-term learning process (the consistent choice of $\pm$ alternatives in the learning process) essentially depends on the initial conditions. If the proportion of pupils who make the correct decision exceeds ρ, then in the end of the learning process, the whole class will do the same. If this percentage of students is less than ρ, then in the behavioral learning process, the average level of intelligence will be reduced to zero.

This law describes the influencing effect of the intellect of individual subjects on the collective mind.

■ In the models $\mathbb{M}\pm$, one of the alternatives is dominant: $N_\pm(k) \to N$ as $k \to \infty$.

3.5. Classification of SEs

In the stochastic case, classification can be done using the *stochastic approximation* (SA) method developed by Robbins and Monro (1951) aimed to find a root of the regression equation and also using the Keifer and Wolfowitz (1952) method for determining the maximum of RF.

The SA procedure is applicable to the SDE that defines the trajectories of normalized SEs using the normalization parameters $a(t_k)$, $t_k := k/N$, $k \geq 0$ (Koroliouk *et al.* 2016a).

3.5.1. *The SA of SEs*

Normalized SEs are considered in the discrete–continuous time $k = [Nt]$, $t \geq 0$, and determined by the solution of the SDEs (see Proposition 3.2.1). The SA procedure is determined by the solution of the SDE:

$$\Delta\alpha_N(t) = a(t)\left[-\frac{V}{N}(\alpha_N(t)) + \sigma(\alpha_N(t))\Delta\mu_N(t+1)\right], \quad t \geq 0. \qquad [3.5.1]$$

The RFI has the form

$$V(c) = V(1-c^2)(c-\rho), \quad |c| \leq 1. \qquad [3.5.2]$$

The stochastic component is characterized by the dispersion $\sigma^2(c)$, $|c| \le 1$, and its first two moments

$$E\Delta\mu_N\left(t+\frac{1}{N}\right) \equiv 0, \quad E\left[\left(\Delta\mu_N\left(t+\frac{1}{N}\right)\right)^2 \,|\, \alpha_N(t)\right] = \frac{1}{N}.$$

The normalization parameters of SA satisfy the following conditions:

$$\sum_{k=1}^{\infty} a(t_k) = \infty, \quad \sum_{k=1}^{\infty} a^2(t_k) < \infty, \quad k \ge 0. \qquad [3.5.3]$$

A normalized SE in discrete–continuous time $t_k = k/N$, $k \ge 1$, is considered with the step $1/N$:

$$\Delta\alpha_N(t) = a(t)\Delta S_N(k+1), \quad t_k \le t < t_{k+1}, \quad k \ge 0.$$

REMARK 3.5.1.– *Discrete Markov diffusion* (DMD) $\alpha_N(t_k)$, $t_k = k/N$, is characterized by the conditional first two moments:

$$E[\Delta\alpha_N(t) \,|\, \alpha_N(t)] = -\frac{a(t)}{N}V(\alpha_N(t)), \qquad [3.5.4]$$

$$E\left[\left[\Delta\alpha_N(t) + \frac{a(t)}{N}V(c)\right]^2 \,|\, \alpha_N(t) = c\right] = \frac{a^2(t)}{N}\sigma^2(c). \qquad [3.5.5]$$

The second moment of DMD increments is calculated using [3.5.4] and [3.5.5]:

$$E[\Delta\alpha_N^2(t) \,|\, \alpha_N(t) = c] = \frac{a^2(t)}{N}B_N(c), \qquad [3.5.6]$$

$$B_N(c) := \sigma^2(c) + \frac{1}{N}V^2(c). \qquad [3.5.7]$$

3.5.2. *Classifiers*

Next, the SA procedure, described in the monograph of Nevelson and Has'minskii (1976) is adapted to SE models.

THEOREM 3.5.1.– (Nevelson and Has'minskii 1976, Theorem 2.7.2). *Let a nonnegative function* $V(c)$, $|c| \le 1$, *exist with zero point* c_0: $V(c_0) = 0$, *which satisfies the inequality*

$$\inf_{|c-c_0|\ge h} [V(c)(c-c_0)] > 0 \quad \textit{for a fixed} \quad h > 0, \qquad [3.5.8]$$

and the second moment of SE increments [3.5.7] *is bounded:*

$$|B_N(c)| = \left|\sigma^2(c) + \frac{V^2(c)}{N}\right| \leq K. \qquad [3.5.9]$$

Then the trajectory of SE $\alpha_N(t)$, $t \geq 0$, converges with probability 1 to the equilibrium point c_0:

$$\alpha_N(t) \xrightarrow{P1} c_0 \ , \ as \quad t \to \infty. \qquad [3.5.10]$$

Theorem 3.5.1 is used to prove the convergence of the SDE solutions [3.5.3] with the use of classifiers that distinguish the points of equilibrium of the RFI [3.5.2] in the classification.

DEFINITION 3.5.1.– The classifiers are given by truncations of the RFIs:

$$V_\rho(c) := V(1 - c^2), \quad |c| \leq 1, \qquad [3.5.11]$$

$$V_{+1}(c) := -V(1 + c)(c - \rho), \quad \rho \leq c \leq 1, \qquad [3.5.12]$$

$$V_{-1}(c) := V(1 - c)(c - \rho), \quad -1 \leq c \leq \rho. \qquad [3.5.13]$$

The classifiers [3.5.11]–[3.5.13] have an essential property: they are strictly separated from zero in the vicinity of the point of equilibrium $c_0 \in \{-1, +\rho\}$ and also have a permanent sign.

The RFI [3.5.2] has the following representation:

$$V(c) = \begin{cases} V_\rho(c)(c - \rho), & |c| \leq 1, \qquad [3.5.11'] \\ -V_{+1}(c)(c - 1), & \rho \leq c \leq +1, \qquad [3.5.12'] \\ V_{-1}(c)(c + 1), & -1 \leq c \leq \rho. \qquad [3.5.13'] \end{cases}$$

Now the SE models are recognized by classifiers [3.5.11]–[3.5.13], taking into account the representations [3.5.11′]–[3.5.13′].

■ Attractive model $\mathbb{M}A : V > 0$, $|\rho| < 1$, is recognized by the classifier, which takes a positive value: $V_\rho(c) = V(1 - c^2) > 0$ for all $c \in (-1, +1)$.

The RFI $V(c)$ is increasing for $c \in (-1, \rho)$ and decreasing for $c \in (\rho, +1)$ according to the formulae [3.5.11′] – [3.5.12′].

This behavior of EPs is classified as the attracting property of the equilibrium point $\rho \in (-1, +1)$.

■ Repulsive model $\mathbb{M}R : V < 0, |\rho| < 1$, is recognized by the classifiers [3.5.12]–[3.5.13] as follows:

$$V_{+1}(c) := -V(1+c)(c-\rho) > 0, \quad c \in (\rho, +1), \qquad [3.5.14]$$

and also

$$V_{-1}(c) := V(1-c)(c-\rho) > 0, \quad c \in (-1, \rho). \qquad [3.5.15]$$

The inequalities [3.5.14] and [3.5.15] mean that the RFI $V(c)$ is increasing for $c \in (\rho, +1)$ up to the value $c_1 = 1$, and decreasing for $c \in (-1, \rho)$ up to the value $c_{-1} = -1$. Such behavior of the corresponding EP is classified as the repulsive property of the equilibrium point $\rho \in (-1, +1)$:

$$P_{\pm}(k) \to \begin{cases} 1, & \text{if } c \in (\rho, +1), \\ -1, & \text{if } c \in (-1, \rho), \end{cases} \quad k \to \infty. \qquad [3.5.16]$$

■ The model $\mathbb{M}A+ : V > 0, \rho \geq +1$ (attractive, dominant $+$) and the model $\mathbb{M}R+ : V < 0, \rho \leq -1$ (repulsive, dominant $+$) are recognized by the classifier [3.5.12] as follows:

$$V_{+1}(c) := -V(1+c)(c-\rho) > 0 \quad \text{for all} \quad c \in (-1, +1). \qquad [3.5.17]$$

For these models, the RFI is used in the following representation:

$$V_0(c) = V_{+1}(c)(c-1) \quad \text{for all} \quad c \in (-1, +1). \qquad [3.5.18]$$

Consequently, the corresponding EPs grow for all $c \in (-1, +1)$, up to the value $+1$ for both models $\mathbb{M}A+$ and $\mathbb{M}R+$.

■ Similarly, we obtain the limit behavior of models $\mathbb{M}A-$ and $\mathbb{M}R-$.

3.5.3. *Classification of SEs*

The classification of SEs, in accordance with the classification of EPs (Proposition 3.4.2), is described in section 3.5.2 and is based on the limit theorems for a normalized stochastic process in the SA scheme.

The classification of behavioral patterns is based on the convergence of almost certainly normalized SE $\alpha_N(t)$.

THEOREM 3.5.2.– (Classification of SEs). *(Proposition 3.4.2) The following almost surely limit results take place:*

- $\mathbb{M}A$: $V > 0$, $|\rho| < 1$:

$$P1 \lim_{t\to\infty} \alpha_N(t) = \rho \quad as \quad t \to \infty. \qquad [3.5.19]$$

- $\mathbb{M}R$: $V < 0$, $|\rho| < 1$:

$$P1 \lim_{t\to\infty} \alpha_N(t) = \begin{cases} -1, & if \quad \alpha_N(0) < \rho, \\ +1, & if \quad \alpha_N(0) > \rho, \end{cases} \quad as \quad t \to \infty. \qquad [3.5.20]$$

- $\mathbb{M}A+$: $V > 0$, $\rho \geq +1$; $\mathbb{M}R+$: $V < 0$, $\rho \leq -1$:

$$P1 \lim_{t\to\infty} \alpha_N(t) = +1 \quad as \quad t \to \infty \quad \forall \alpha_N(0) : \quad |\alpha_N(0)| < 1. \qquad [3.5.21]$$

- $\mathbb{M}A-$: $V > 0$, $\rho \leq -1$; $\mathbb{M}R-$: $V < 0$, $\rho \geq +1$:

$$P1 \lim_{t\to\infty} \alpha_N(t) = -1 \quad as \quad t \to \infty \quad \forall \alpha_N(0) \ : \quad |\alpha_N(0)| < 1. \qquad [3.5.22]$$

PROOF 3.43.– The generator of the DMP $\alpha_N(t)$ is used:

$$L_N\Phi(c) = NE[\Phi(c + \Delta\alpha_N(t)) - \Phi(c) \,|\, \alpha_N(t) = c]. \qquad [3.5.23]$$

LEMMA 3.5.1.– *On the test functions*

$$\Phi(c) := (c - c_0)^2, \quad |c| \leq 1, \quad c_0 \in \{-1, +1, \rho\}, \qquad [3.5.24]$$

the generator [3.5.23] *has the following representation:*

$$L_N(c - c_0)^2 = -2a(t)V_{c_0}(c)(c - c_0)^2 + a^2(t)B_N(c), \qquad [3.5.25]$$

where

$$B_N(c) := \sigma^2(c) + \frac{V^2(c)}{N}, \qquad [3.5.26]$$

Using Lemma 3.5.1, the main condition [3.5.8] of Theorem 3.4.1 transforms into the following equality:

$$V(c)(c - c_0) = \frac{1}{4}V_{c_0}(c)(c - c_0)^2, \quad c \in \{-1, +1, \rho\}. \qquad [3.5.27]$$

Hence, by Theorem 3.5.1, the stochastic process $\alpha_N(t)$, $t \geq 0$, characterized by generators [3.5.23]–[3.5.26], converges with probability 1, as $t \to \infty$, to the equilibrium point $c_0 \in \{-1, +1, \rho\}$. The convergences [3.5.19]–[3.5.22] of Theorem 3.5.2 are proved. □

REMARK 3.5.2.– In the MSE models with more than two alternatives, the behavioral patterns become more complex. However, in essence, the main features are always the same: after a fairly large number of stages, the frequencies in the equilibrium state are characterized by three types of equilibrium: attracting, repulsive and absorbing.

In processes with a large number of alternatives, the behavior of the equilibrium state (in the space of parameters) can be a point, a line, a plane, and so on, up to an R-dimensional subspace.

3.6. Evolutionary model of ternary SEs

On the example of ternary SEs, we will demonstrate the analysis of complexity of the MSEs. Only the classification of EPs, taking into account the representation of their trajectories by the DEE, is considered with use of the RFIs, which depend only on the fluctuations of the components of EPs relative to the stationary state (equilibrium).

3.6.1. *Basic assumptions*

Let us consider SEs with a main alternative A_0 and two additional alternatives A_1 and A_2.

The main idea is to choose the main factor that essentially determines the state of the SE, supplemented by additional alternatives in such a way that the aggregation of the main factor and its alternatives completely describe the temporal dynamics of SE.

The *basic characteristic* of the main and additional alternatives are their probabilities (frequencies): P_0 of the main factor and P_1 and P_2 of the additional alternatives, for which the balance condition takes place:

$$P_0 + P_1 + P_2 = 1. \qquad [3.6.1]$$

The dynamics of EPs is determined by a *linear regression function*, which defines the EP values in the next stage of observation, depending on the value at the current stage.

Consider the sequence of EP values, which depend on the observation stage, or on the discrete time parameter $k \geq 0$, which is the same:

$$P(k) := (P_0(k), P_1(k), P_2(k)), \quad k \geq 0,$$

as well as their increments at kth time instant:

$$\Delta P(k+1) := P(k+1) - P(k), \quad k \geq 0.$$

The linear RFI is determined by a matrix generated by drift parameters in the following DEE:

$$\Delta P(k+1) = -\widehat{\mathbb{V}} P(k), \quad k \geq 0, \tag{3.6.2}$$

where

$$\widehat{\mathbb{V}} := [\widehat{V}_{mn},\ 0 \leq m, n \leq 2], \tag{3.6.3}$$

$$\widehat{V}_{mm} = 2V_m, \quad \widehat{V}_{mn} = -V_n, \quad 0 \leq n \leq 2, \quad n \neq m. \tag{3.6.4}$$

The drift parameters V_0, V_1 and V_2 satisfy the following inequality:

$$|V_m| \leq 1, \quad 0 \leq m \leq 2. \tag{3.6.5}$$

An important feature of the EP is the presence of stationary state ρ (equilibrium), defined by zero of the RFI:

$$\widehat{\mathbb{V}} \rho = 0, \tag{3.6.6}$$

or in scalar form

$$\widehat{V}_m \rho := \widehat{V}_{m0}\, \rho_0 + \widehat{V}_{m1}\, \rho_1 + \widehat{V}_{m2}\, \rho_2 = 0, \quad 0 \leq m \leq 2. \tag{3.6.7}$$

Of course, there is a balance condition

$$\rho_0 + \rho_1 + \rho_2 = 1. \tag{3.6.8}$$

REMARK 3.6.1.– Given equations [3.6.6] and [3.6.7] and the balance condition [3.6.8], we have the explicit formulae for the equilibria:

$$\rho_m = \frac{V_m^{-1}}{\overline{V}}, \quad 0 \leq m \leq 2, \tag{3.6.9}$$

$$\overline{V} := V_0^{-1} + V_1^{-1} + V_2^{-1}, \tag{3.6.10}$$

or equivalently

$$\rho_0 = \frac{V_1 V_2}{\overline{V}}, \quad \rho_1 = \frac{V_0 V_2}{\overline{V}}, \quad \rho_2 = \frac{V_0 V_1}{\overline{V}}, \tag{3.6.8'}$$

$$\overline{V} := V_1 V_2 + V_0 V_2 + V_0 V_1. \tag{3.6.11}$$

The validity of formulae [3.6.9] and [3.6.8′] can be easily verified by substituting them into equations [3.6.7] and [3.6.8]. Obviously, there is the limitation $V \neq 0$.

Now, the probabilities of the fluctuations with respect to equilibrium

$$\widehat{P}_m(k) := P_m(k) - \rho_m, \quad 0 \leq m \leq 2, \qquad [3.6.12]$$

satisfy the DEE [3.6.2].

Basic assumption. The dynamics of the EP is determined by the DEE for the probabilities of alternatives $\widehat{P}_0(k)$, $\widehat{P}_1(k)$ and $\widehat{P}_2(k)$:

$$\Delta \widehat{P}(k+1) = -\widehat{\mathbb{V}}\widehat{P}(k), \quad k \geq 0, \qquad [3.6.13]$$

or in a scalar form

$$\Delta \widehat{P}_m(k+1) = \widehat{V}_{m0}\, \widehat{P}_0(k) + \widehat{V}_{m1}\, \widehat{P}_1(k) + \widehat{V}_{m2}\, \widehat{P}_2(k),$$
$$0 \leq m \leq 2, \quad k \geq 0. \qquad [3.6.14]$$

In addition, the initial values should be given as

$$\widehat{P}(0) = (\widehat{P}_0(0), \widehat{P}_1(0), \widehat{P}_2(0)).$$

REMARK 3.6.2.– Determination of dynamics using the linear RFs [3.6.13] and [3.6.14] in the evolutionary model of ternary SEs, with additional constraints:

$$0 \leq P_m(k) \leq 1, \quad 0 \leq m \leq 2, \quad k \geq 0, \quad 0 \leq \rho_m \leq 1, \quad 0 \leq m \leq 2,$$

have, as a consequence, the balance condition [3.6.1] and the equilibrium definition [3.6.6] for the solutions of the difference equations [3.6.2] and [3.6.3].

3.6.2. *The model interpretation and analysis*

The construction of the EP model is carried out in several stages. First, the main factor, which is characterized by probability (or frequency, concentration, etc.) should be chosen. In this case, the additional alternatives are identified, as well as their probabilities.

The probabilities of the main factor and of the additional alternatives satisfy the balance condition [3.6.1] or, equivalent, the condition [3.6.8].

The main alternative probability dynamics P_0, as well as the additional alternatives P_1 and P_2, given by the following difference equations for the probability of the fluctuations for all $k \geq 0$:

$$\begin{aligned}
\Delta\widehat{P}_0(k+1) &= V_1\widehat{P}_1(k) + V_2\widehat{P}_2(k) - 2V_0\widehat{P}_0(k), \\
\Delta\widehat{P}_1(k+1) &= V_0\widehat{P}_0(k) + V_2\widehat{P}_2(k) - 2V_1\widehat{P}_1(k), \\
\Delta\widehat{P}_2(k+1) &= V_0\widehat{P}_0(k) + V_1\widehat{P}_1(k) - 2V_2\widehat{P}_2(k).
\end{aligned} \qquad [3.6.15]$$

The increments of the fluctuations of the main and additional factors

$$\Delta\widehat{P}_m(k+1) := \widehat{P}_m(k+1) - \widehat{P}_m(k), \quad 0 \leq m \leq 2, \quad k \geq 0,$$

are determined by the values of the drift parameters V_0, V_1 and V_2.

REMARK 3.6.3.– The probabilities of the fluctuations in [3.6.12] satisfy the balance condition:

$$\widehat{P}_0(k) + \widehat{P}_1(k) + \widehat{P}_2(k) = 0, \quad k \geq 0, \qquad [3.6.16]$$

and according to formula [3.6.12], it is obvious that

$$\Delta\widehat{P}_m(k) = \Delta P_m(k), \quad 0 \leq m \leq 2, \quad k \geq 0. \qquad [3.6.17]$$

Equation [3.6.15] characterizes two basic principles of interaction of alternatives: *stimulation* (positive terms) and *deterrence* (negative terms).

The existence of an equilibrium point for the fluctuation increment of the RF [3.6.6] makes it possible to analyze the dynamics of the EP (as $k \to \infty$) taking into account the possible values of the drift parameters that satisfy the restrictions [3.6.5].

The dynamics of the main factor probability is described by the *following models*.

PROPOSITION 3.6.1.– The main factor probabilities $P_0(k)$, $k \geq 0$, determined by the solution of the difference equation [3.6.15], as well as the basic assumption [3.6.13] of equilibrium [3.6.14], evolves, as $k \to \infty$, to the following behavioral patterns:

■ $\mathbb{M}A$: $V_0 > 0,\ 0 < \rho_0 < 1$:

$$\lim_{k\to\infty} P_0(k) = \rho_0. \qquad [3.6.18]$$

■ $\mathbb{M}R$: $V_0 < 0,\ 0 < \rho_0 < 1$:

$$\lim_{k\to\infty} P_0(k) = \begin{cases} 1 & \text{if} \quad P_0(0) > \rho_0, \\ 0 & \text{if} \quad P_0(0) < \rho_0. \end{cases} \qquad [3.6.19]$$

■ $\mathbb{M}+$: $\rho_0 \notin (0,1),\ V_0 < 0$:

$$\lim_{k\to\infty} P_0(k) = 1. \qquad [3.6.20]$$

■ $\mathbb{M}-$: $\rho_0 \notin (0,1),\ V_0 > 0$:

$$\lim_{k\to\infty} P_0(k) = 0. \qquad [3.6.21]$$

REMARK 3.6.4.– Of course, the basic models of the dynamics can be formulated in terms of values of drift parameter V_0 (see Table 3.1).

	$V_0 > 0$	$V_0 < 0$
$0 < \rho_0 < 1$	Attractive equilibrium: $P_0(k) \to \rho_0, \quad k \to \infty$	Repulsive equilibrium: $P_0(k) \to \begin{cases} 1, & \text{if } P_0(0) > \rho_0, \\ 0, & \text{if } P_0(0) < \rho_0, \end{cases}$ $k \to \infty$
$\rho_0 > 1$	Repulsive domination $-$: $P_0(k) \to 0, \quad k \to \infty$	Attractive domination $+$: $P_0(k) \to 1, \quad k \to \infty$
$\rho_0 < 1$	Attractive domination \$-\$: $P_0(k) \to 0, \quad k \to \infty$	Repulsive domination $+$: $P_0(k) \to 1, \quad k \to \infty$

Table 3.1. *Table of models*

REMARK 3.6.5.– The same models of SE dynamics can be formulated, taking into account the parameters V_1, ρ_1 or V_2, ρ_2.

3.7. Equilibrium states in the dynamics of ternary SEs

In section 3.6, the model of *ternary SEs* with persistent linear regression and the presence of equilibrium states is considered in terms of the main factor dynamics.

Hereafter, the ternary model is considered in a deeper scheme of multivariate discrete stationary Markov diffusion, which is determined by vector SDEs.

3.7.1. *Building a model*

Ternary statistical experiments (TSEs) have three alternatives $0 \leq m \leq 2$:

$$S_N^{(m)}(k) := \frac{1}{N}\sum_{n=1}^{N} \delta_n^{(m)}(k),$$

$$\delta_n^{(m)}(k) = \begin{cases} 1, & \text{if } n\text{th sample contains } m\text{th alternative,} \\ 0, & \text{otherwise.} \end{cases}$$

That is,

$$S_N(k) := (S_N^{(0)}(k), S_N^{(1)}(k), S_N^{(2)}(k)) \quad : \\ : \quad S_N^{(0)}(k) + S_N^{(1)}(k) + S_N^{(2)}(k) = 1, \quad k \geq 0.$$

The *preliminary assumption* at the RFI is

$$E[\Delta S_N(k) \mid S_N(k)] = -\mathbb{V} S_N(k), \quad k \geq 0, \\ \Delta S_N(k) := S_N(k+1) - S_N(k). \qquad [3.7.1]$$

Now consider the *martingale difference* (stochastic component)

$$\Delta\mu_N(k) = \Delta S_N(k) - E[\Delta S_N(k) \mid S_N(k)], \\ E[\Delta\mu_N(k)] = 0, \quad E[\Delta\mu_N(k)]^2 = \sigma_N^2. \qquad [3.7.2]$$

Then *the dynamics of TSE* is described by the SDE:

$$\Delta S_N(k) = -\mathbb{V} S_N(k) + \Delta\mu_N(k), \quad k \geq 0, \qquad [3.7.3]$$

where

$$\mathbb{V} = \begin{bmatrix} 2V_0 & -V_1 & -V_2 \\ -V_0 & 2V_1 & -V_2 \\ -V_0 & -V_1 & 2V_2 \end{bmatrix}, \quad |V_m| \leq 1, \quad 0 \leq m \leq 2. \qquad [3.7.4]$$

In this case, the balance condition is

$$\vec{1}\mathbb{V}\rho = 0. \qquad [3.7.5]$$

3.7.2. *The equilibrium state and fluctuations*

As before, the equilibrium state is defined as zero of the matrix RFI

$$\widehat{\mathbb{V}}\rho = 0, \qquad [3.7.6]$$

or in a scalar form

$$\widehat{V}^{(m)}\rho := \widehat{V}_0^{(m)}\rho_0 + \widehat{V}_1^{(m)}\rho_1 + \widehat{V}_2^{(m)}\rho_2 = 0, \quad 0 \leq m \leq 2. \qquad [3.7.7]$$

The balance condition is

$$\rho_0 + \rho_1 + \rho_2 = 1. \qquad [3.7.8]$$

REMARK 3.7.1.– We have the explicit formulae for the equilibria:

$$\rho_m = \frac{V_m^{-1}}{\overline{V}}, \quad 0 \leq m \leq 2, \quad \overline{V} := V_0^{-1} + V_1^{-1} + V_2^{-1}, \qquad [3.7.9]$$

or equivalently,

$$\rho_0 = \frac{V_1 V_2}{\overline{V}}, \quad \rho_1 = \frac{V_0 V_2}{\overline{V}}, \quad \rho_2 = \frac{V_0 V_1}{\overline{V}},$$
$$\overline{V} := V_1 V_2 + V_0 V_2 + V_0 V_1, \quad V \neq 0. \qquad [3.7.10]$$

As before, the fluctuations with respect to equilibrium are defined as

$$\widehat{S}_N(k) := S_N(k) - \rho. \qquad [3.7.11]$$

Basic assumption. The dynamics of SEs is given by the following SDE:

$$\Delta\widehat{S}_N(k) = -\widehat{\mathbb{V}}\widehat{S}_N(k) + \Delta\mu_N(k+1), \quad k \geq 0, \qquad [3.7.12]$$

or in scalar form

$$\Delta\widehat{S}_N^{(m)}(k) = \widehat{V}_0^{(m)}\,\widehat{S}_1^{(m)}(k) + \widehat{V}_1^{(m)}\,\widehat{S}_1^{(m)}(k) + \widehat{V}_2^{(m)}\,\widehat{S}_2^{(m)}(k)$$
$$+ \Delta\mu_N^{(m)}(k+1), \quad 0 \leq m \leq 2, \quad k \geq 0. \qquad [3.7.13]$$

3.7.3. *Classification of TSEs*

By virtue of the basic assumption [3.7.12] and the existence of the equilibria [3.7.9] and [3.7.10], it is possible to analyze the asymptotic dynamics of TSEs.

The classification of limit dynamics of TSEs is realized on the basis of the law of large numbers by means of a double limit passage in terms of the sample size and the discrete time parameter.

PROPOSITION 3.7.1.– The frequency dynamics of the main factor $S_{0N}(k)$, $k \geq 0$, as $k \to \infty$ and $N \to \infty$, is determined by *several scenarios*. The convergence is intended in the mean-square.

- $\mathbb{M}A$: $V_0 > 0,\ 0 < \rho_0 < 1$:

$$\lim_{\substack{k\to\infty \\ N\to\infty}} E[S_{0N}(k) - \rho_0]^2 = 0. \tag{3.7.14}$$

- $\mathbb{M}R$: $V_0 < 0,\ 0 < \rho_0 < 1$:

$$\lim_{\substack{k\to\infty \\ N\to\infty}} E[S_N^{(0)}(k) - \delta(S_N^{(0)}(0))]^2 = 0, \qquad \delta(c) := \begin{cases} 0 & c < \rho_0, \\ 1 & c > \rho_0. \end{cases} \tag{3.7.15}$$

- $\mathbb{M}+$: $\rho_0 \notin (0,1),\ V_0 < 0$:

$$\lim_{\substack{k\to\infty \\ N\to\infty}} E[S_N^{(0)}(k) - 1]^2 = 0. \tag{3.7.16}$$

- $\mathbb{M}-$: $\rho_0 \notin (0,1),\ V_0 > 0$:

$$\lim_{\substack{k\to\infty \\ N\to\infty}} E[S_N^{(0)} - 0]^2 = 0. \tag{3.7.17}$$

Now we can summarize the classification results of TSEs in the form of Table 3.2.

	$V_0 > 0$	$V_0 < 0$
$0 < \rho_0 < 1$	Attractive equilibrium: $S_{0N}(k) \to \rho_0$ as $N \to \infty, k \to \infty$	Repulsive equilibrium: $S_{0N}(k) \to \begin{cases} 1, & S_{0N}(0) > \rho_0, \\ 0, & S_{0N}(0) < \rho_0. \end{cases}$ as $N \to \infty, k \to \infty$
$\rho_0 > 1$	Repulsive degradation – $P_0(k) \to 0$ as $N \to \infty, k \to \infty$	Attractive domination + $P_0(k) \to 1$ as $N \to \infty, k \to \infty$
$\rho_0 < 0$	Attractive degradation – $S_{0N}(k) \to 0$ as $N \to \infty, k \to \infty$	Repulsive domination + $S_{0N}(k) \to 1$ as $N \to \infty, k \to \infty$

Table 3.2. *Classification of TSE as* $N,\ k \to \infty$

4

Modeling and Numerical Analysis of Statistical Experiments

This statistical experiments (SE) trajectories and verification of statistical estimations of SE parameters. In order to illustrate modeling of interaction of biological macromolecules, a discrete Markov diffusion (DMD) model with the corresponding correlation statistics is used.

4.1. Numerical verification of generic model

The generic model of SEs is given by averaged sums of sampling random variables. An essential feature of SEs is that the probabilities of the values at the next stage of observation depend only on the averaged value of SEs at the previous stage.

The dynamics of the probabilities of sampling variables is generated by the regression functions of increments (RFIs; linear or nonlinear).

4.1.1. *Evolutionary processes with linear and nonlinear RFIs*

The dynamic models of SEs are defined as averaged sums

$$S_N(k) := \frac{1}{N}\sum_{n=1}^{N} \delta_n(k), \quad k \geq 0, \qquad [4.1.1]$$

of sampling random variables $\delta_n(k)$, $1 \leq n \leq N$, taking a finite number of values, equally distributed at different n, $1 \leq n \leq N$, and mutually independent at fixed $k \geq 0$ (see Chapter 1).

The main objective of the study is a *binary SE*, defined by sampling random variables $\delta_n(k)$, $1 \leq n \leq N$, which take two values ± 1.

The binary and frequency SEs are connected as follows (section 1.1):

$$S_N(k) = S_N^+(k) - S_N^-(k), \quad k \geq 0. \qquad [4.1.2]$$

The frequency SEs are given by averaged sums:

$$S_N^\pm(k) := \frac{1}{N}\sum_{n=1}^{N} \delta_n^\pm(k), \quad \delta_n^\pm(k) := I\{\delta_n(k) = \pm 1\}, \quad k \geq 0. \qquad [4.1.3]$$

Here $I(A)$ is an indicator of a random event A.

The obvious identity

$$S_N^+(k) + S_N^-(k) \equiv 1 \quad \forall k \geq 0$$

provides a one-to-one correspondence between binary and frequency SEs:

$$S_N^\pm(k) = \frac{1}{2}[1 \pm S_N(k)], \; S_N(k) = 2S_N^+(k) - 1 = 1 - 2S_N^-(k), \; k \geq 0. \qquad [4.1.4]$$

So, without generality limiting, only positive SEs can be considered:

$$S_N^+(k) := \frac{1}{N}\sum_{n=1}^{N} \delta_n^+(k), \quad \delta_n^+(k) := I\{\delta_n(k) = +1\}, \quad k \geq 0. \qquad [4.1.5]$$

However, it may be useful to compare the simulation results of the frequency [4.1.5] and binary SEs [4.1.1], taking into account the connection [4.1.4].

The dynamics of frequency [4.1.5] is determined by evolutionary processes (EPs) $P_\pm(k)$, $k \geq 0$, given by the conditional mathematical expectations:

$$P_\pm(k+1) = E[\delta_n^\pm(k+1) \,|\, S_N^\pm(k) = P_\pm(k)], \quad 1 \leq n \leq N, \; k \geq 0. \qquad [4.1.6]$$

Otherwise (see section 1.1),

$$P_\pm(k+1) = E[S_N^\pm(k+1) \,|\, S_N^\pm(k) = P_\pm(k)], \quad k \geq 0. \qquad [4.1.7]$$

Obviously, the EPs [4.1.7] do not depend on the volume N of the sample.

The dynamics of binary SEs [4.1.1] is also determined by the binary EP $C(k)$, $k \geq 0$, given by the conditional mathematical expectations

$$C(k+1) := E[\delta_n(k+1) \mid S_N(k) = C(k)], \quad 1 \leq n \leq N, \quad k \geq 0, \quad [4.1.8]$$

or, equivalently,

$$C(k+1) := E[S_N(k+1) \mid S_N(k) = C(k)], \quad k \geq 0. \quad [4.1.9]$$

So EPs—frequency EPs $P_\pm(k)$, $k \geq 0$, and binary EPs $C(k)$, $k \geq 0$—are given by the conditional mathematical expectations [4.1.7] and [4.1.9], respectively.

The linear RFI defines the fluctuations of EPs with respect to the equilibria:

$$V_0(p_\pm) = -V_0(p_\pm - \rho_\pm). \quad [4.1.10]$$

The nonlinear RFI also specifies the fluctuations of EPs in relation to the equilibrium with additional multipliers:

$$V_0(p_\pm) = -V_0 \cdot p_+ p_-(p_\pm - \rho_\pm). \quad [4.1.11]$$

The presence of the equilibrium state in the linear component of RFI [4.1.11] ensures the fulfillment of the global balance condition:

$$V_0(p_+) + V_0(p_-) \equiv 0, \quad p_+ + p_- = 1, \quad \rho_+ + \rho_- = 1. \quad [4.1.12]$$

The dynamics of frequency EPs [4.1.7] are given by the RFI [4.1.12]:

$$\Delta P_\pm(k+1) = -V_0(P_\pm(k) - \rho_\pm), \quad k \geq 0, \quad [4.1.13]$$

expressed in terms of the fluctuations

$$\widehat{p}_\pm := p_\pm - \rho_\pm, \quad p_\pm = \widehat{p}_\pm + \rho_\pm. \quad [4.1.14]$$

Equation [4.1.13] means that the conditional mathematical expectations [4.1.7] and [4.1.9] are given by RFIs, which reflect the basic principle of the interaction of elementary random events: the principle of "stimulation-deterrence."

Then the dynamics of EPs $\widehat{P}_+(k)$, $k \geq 0$, is determined by a *difference evolutionary equation* (DEE) for fluctuations with a linear RFI [4.1.10]:

$$\Delta \widehat{P}_\pm(k+1) = -V \widehat{P}_\pm(k), \quad k \geq 0, \quad [4.1.15]$$

or with a nonlinear RFI [4.1.11]:

$$\Delta\widehat{P}_{\pm}(k+1) = -(\rho_+ + \widehat{P}_+(k))(\rho_- + \widehat{P}_-(k))V\widehat{P}_{\pm}(k), \quad k \geq 0. \quad [4.1.16]$$

Similarly, the binary EP fluctuations are determined as follows:

$$\widehat{C}(k) := C(k) - \rho, \quad \rho = \rho_+ - \rho_-, \quad k \geq 0, \quad [4.1.17]$$

with a linear RFI [4.1.10]

$$\Delta\widehat{C}(k+1) = -V\widehat{C}(k), \quad k \geq 0, \quad [4.1.18]$$

or with a nonlinear RFI [4.1.11]

$$\Delta\widehat{C}(k+1) = -\frac{1}{4}V\left[1 - C^2(k)\right]\widehat{C}(k), \quad k \geq 0. \quad [4.1.19]$$

The DEEs [4.1.15] and [4.1.18] realize an essential feature of the interaction of elements of complex stochastic systems: the dynamics of EPs [4.1.7] and [4.1.9] at $(k+1)$th stage is determined by averaged values of the SEs in the previous step k.

4.1.2. *Generic model of trajectory generation*

Let us fix the couple of basic parameters of an SE (V, ρ).

In order to generate the samples of pseudo-random values that simulate the trajectories of a static experiment

$$S_N(k) := \frac{1}{N}\sum_{n=1}^{N}\delta_n(k), \quad -1 \leq S_N(k) \leq 1, \quad k \geq 0, \quad [4.1.20]$$

where $\delta_n(k)$, $1 \leq n \leq N$, $k \geq 0$, are independent Bernoulli pseudo-random numbers with distribution

$$\delta_n(k) = \begin{cases} 1, & \text{with probability} \quad P_+(k), \\ -1, & \text{with probability} \quad 1 - P_+(k), \end{cases} \quad k \geq 0, \quad [4.1.21]$$

we must gradually calculate the probabilities $P_+(k)$, $k \geq 0$, which, according to Chapter 1, are the trajectory of the frequency EP [1.1.35] (see Chapter 1), determined by the DEE

$$\Delta P_+(k+1) = -V\left[P_+(k) - \rho_+\right], \quad k \geq 0, \quad [4.1.22]$$

which provides a basis for recurrent calculations of the probabilities $P_+(k)$, $k \geq 1$, and the subsequent generation of Bernoulli binary random variables [4.1.2].

Here the frequency equilibrium is

$$\rho_+ = \frac{1}{2}[1 + \rho]. \qquad [4.1.23]$$

Thus, the algorithm of numerical simulation of the generic model of SEs can be described as follows:

===

Algorithm A1

The following parameters are specified:
N (sample size), K (number of stages);
V (value of the drift parameter), ρ (equilibrium value);
$P_+(0)$ (initial value of probability);

==========

• the calculation of ρ_+ by formula [4.1.4];
■ for k from 1 to K, carry out the following procedure:
► calculate the probability $P_+(k)$ (see [4.1.3]) by the formula

$$P_+(k+1) = (1 - V)P_+(k) + V\rho_+;$$

► Generate N of uniformly distributed random values $u_n \in U(0,1)$, $1 \leq n \leq N$;
► build a binary random sample by the rule

$$\delta_n(k) = \begin{cases} 1, & \text{if} \quad u_n < P_+(k), \\ -1, & \text{if} \quad u_n \geq P_+(k), \end{cases}$$

► calculate $S_N(k)$ by formula [4.1.1];
► return to ■ to complete the loop by k.

==========

Thus, for the given parameters N, K, V, ρ and $P_+(0)$, a sequence, by k, is constructed for numerical simulation of the trajectory of an SE $S_N(1), S_N(2), ..., S_N(K)$.

Figures 4.1–4.4 show the graphs of the generic fluctuation SEs:

$$\widehat{S}_N(1), \widehat{S}_N(2), ..., \widehat{S}_N(K), \quad \widehat{S}_N(k) := S_N(k) - \rho,$$

for $N = 220$, $K = 400$, $V = 0,8$ or $V = 1,95$.

The generic SE model simulation of trajectories shows changes in stationary processes with increasing the value of parameter V. So, for $V = 1,95$ (close to the extreme value $V = 2$, Figure 4.4(a)), the essential fluctuations are observed. In addition, the EP is stabilized at the value of equilibrium much slower for $V = 1,95$, than for $V = 0,8$.

4.2. Numerical verification of DMD

The simulation of DMD trajectories is more technically feasible than the numerical modeling of the generic model, given in section 4.1. The reason is the possibility of creating structured computational procedures that are self-replicating and easily expanded, if necessary, without modifying the existing computing blocks.

As a result, the efficiency of calculations increases by one or two orders of magnitude, which allows us to operate millions of samples of numerical simulation results or to process millions (and in the long run, billions) of real experimental data using economic computing resources.

4.2.1. *Simulation of DMD trajectories*

The DMD model is the solution of the stochastic difference equation (SDE):

$$\Delta\alpha(k+1) = -V\alpha(k) + \sigma\Delta\mu(k+1), \quad k \geq 0, \quad \alpha(0) \text{ is fixed.} \quad [4.2.1]$$

The trajectories of two-component DMDs $(\alpha(k), \Delta\alpha(k+1))$, $k \geq 0$, are simulated by the following procedure.

==

Algorithm A2

=== Initial data ===

Specified settings:

K (number of stages);

V (the value of the drift parameter);

σ (the dispersion of the stochastic component $\sigma\Delta\mu(k+1)$);

$\alpha(0)$ (the initial value of DMD);

=== The computational procedure ===

■ for k from 1 to K, carry out the following procedure:

• generation of a uniformly distributed random number $Unif.[0,1]$;

• transformation $Unif.[0,1] \rightsquigarrow N(0,\sigma^2)$ (generation of a normally distributed stochastic component $\sigma\Delta\mu(k+1)$);

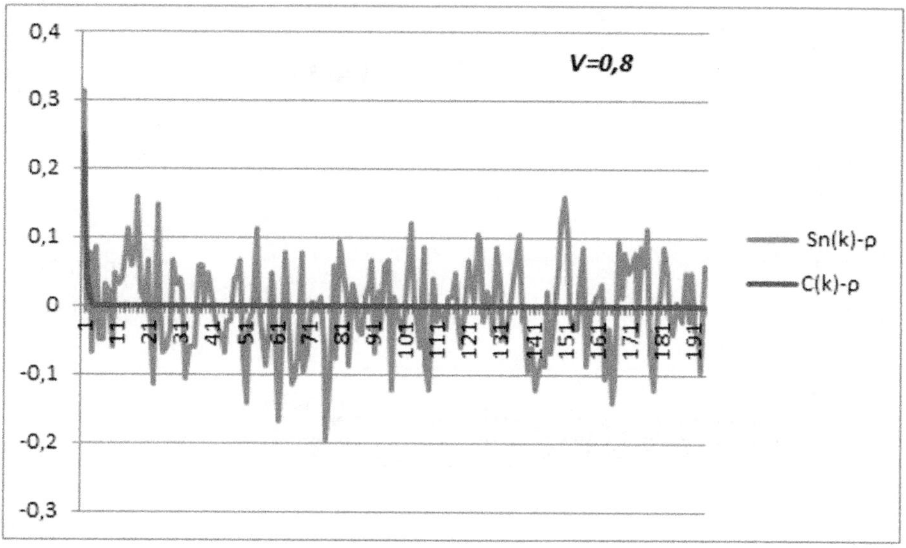

(a)

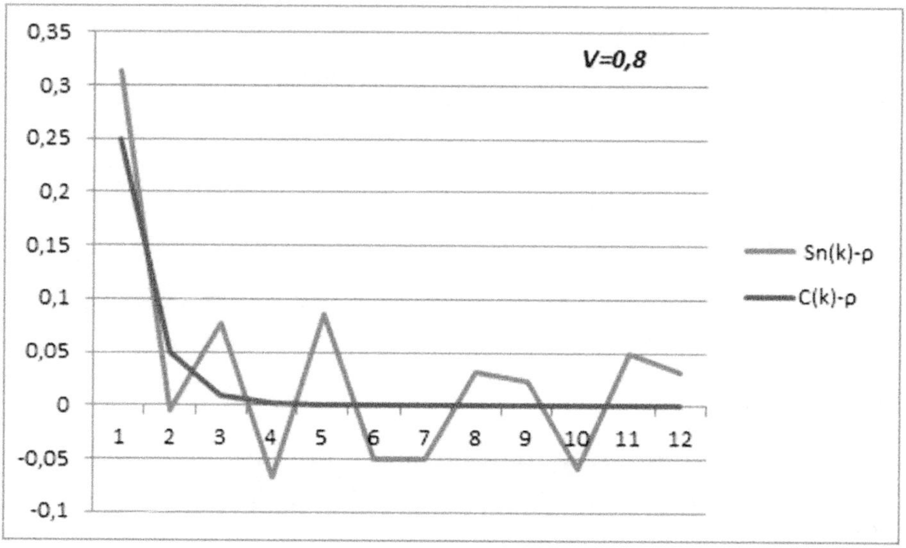

(b)

Figure 4.1. *Fluctuations of binary SEs at* $V = 0,8$ *at different time intervals: (a)* $\widehat{S}_N(k)$ *and* $\widehat{C}(k)$ *for* $1 \leq k \leq 200$*; (b)* $\widehat{S}_N(k)$ *and* $\widehat{C}(k)$ *for* $1 \leq k \leq 12$*. For a color version of the figures in this book, see http://www.iste.co.uk/koroliouk/dynamics.zip*

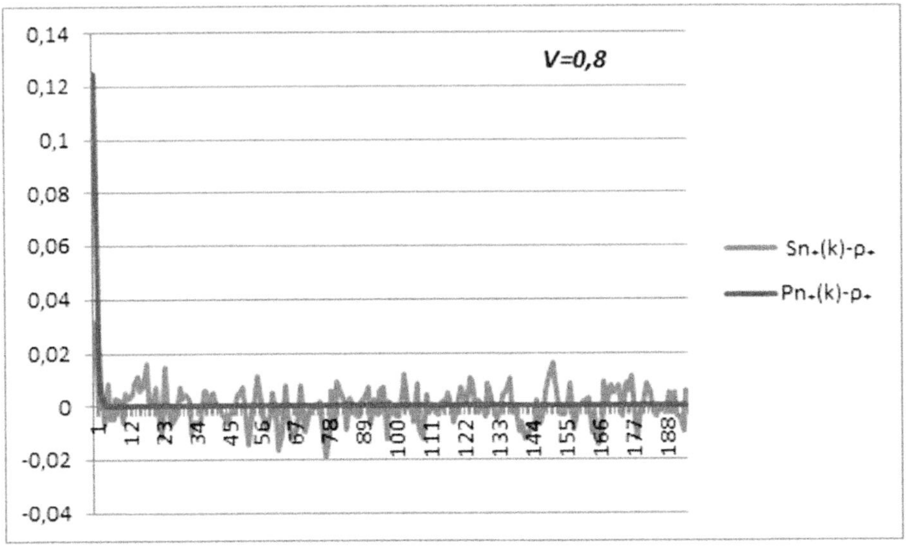

(a)

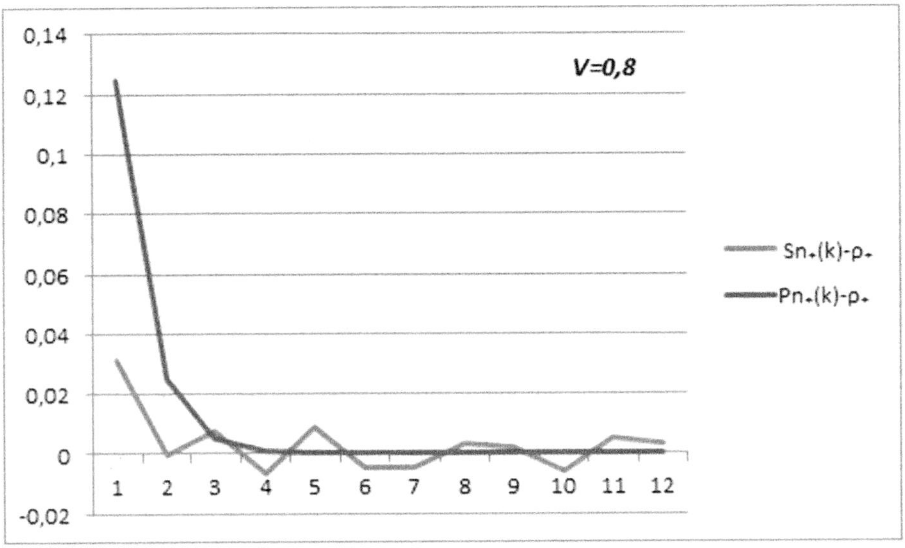

(b)

Figure 4.2. *Fluctuations of binary SEs at* $V = 0,8$ *at different time intervals: (a)* $\widehat{S}_N^+(k)$ *and* $\widehat{P}^+(k)$ *for* $1 \le k \le 200$*; (b)* $\widehat{S}_N^+(k)$ *and* $\widehat{P}^+(k)$ *for* $1 \le k \le 12$

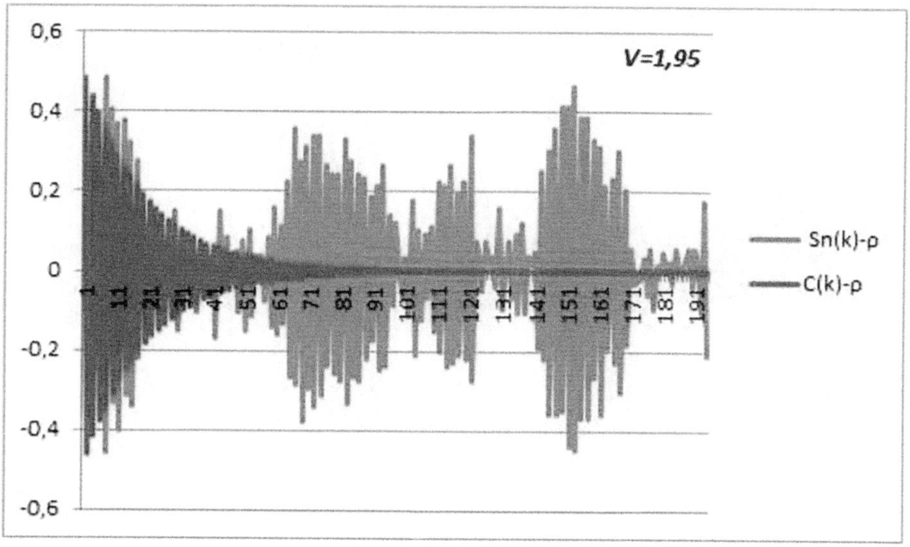

(a)

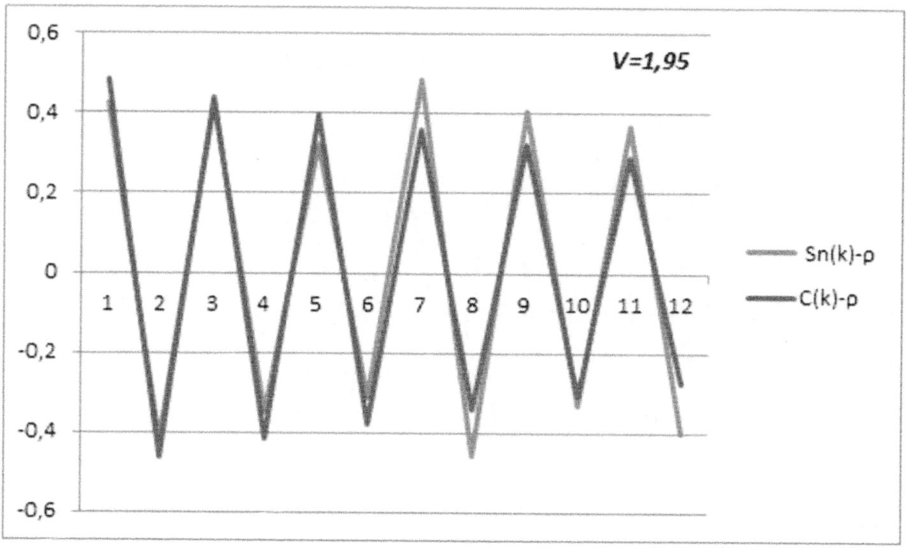

(b)

Figure 4.3. *Fluctuations of binary SEs at* $V = 1,95$ *at different time intervals: (a)* $\widehat{S}_N(k)$ *and* $\widehat{C}(k)$ *for* $1 \leq k \leq 200$*; (b)* $\widehat{S}_N(k)$ *and* $\widehat{C}(k)$ *for* $1 \leq k \leq 12$

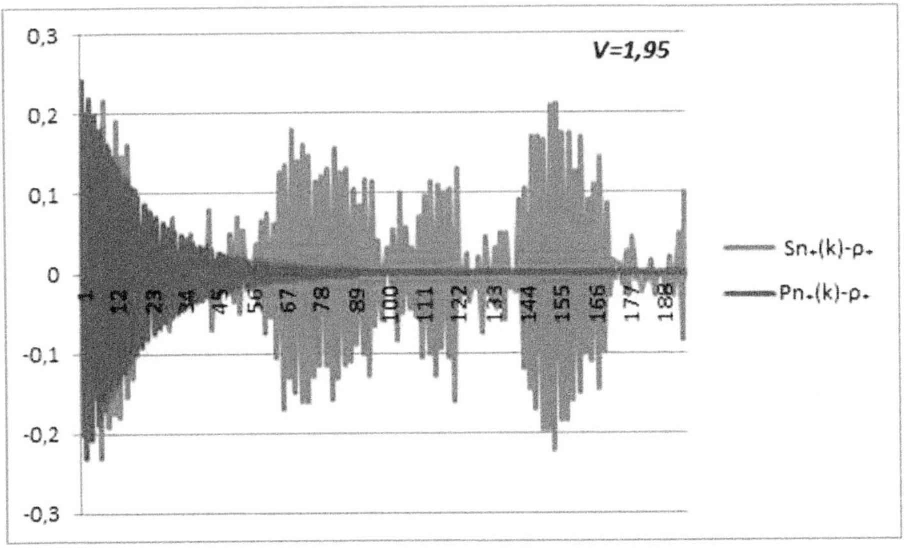

(a)

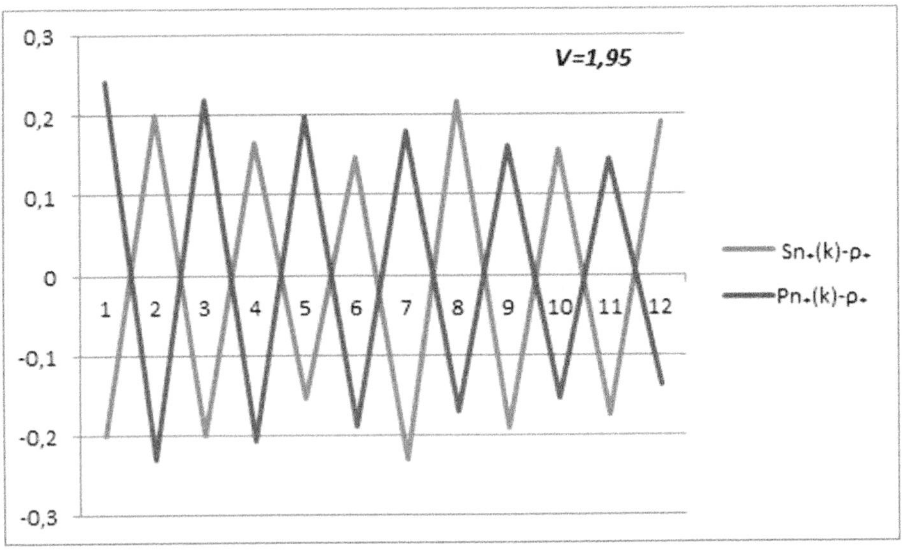

(b)

Figure 4.4. *Fluctuations of binary SEs at $V = 1,95$ at different time intervals: (a) $\widehat{S}_N^+(k)$ and $\widehat{P}^+(k)$ for $1 \leq k \leq 200$; (b) $\widehat{S}_N^+(k)$ and $\widehat{P}^+(k)$ for $1 \leq k \leq 12$*

• calculate recursively $\alpha(k)$, $\Delta\alpha(k+1)$ by formula (4.2.1), taking into account the initial value $\alpha(0)$;
• calculate the covariance statistics $\alpha^2(k)$, $[\Delta\alpha(k+1)]^2$, $\alpha(k)\Delta\alpha(k+1)$.
==========

Thus, for the given parameters K, V, σ and $\alpha(0)$, a sample sequence is constructed by k numerical simulations of the points of DMD trajectory $\alpha(1), \alpha(2), ..., \alpha(K)$, considered as an SE.

4.2.2. *Estimation of DMD parameters*

As shown in section 3.1.2, the estimates of basic parameters of the generic model SE can be calculated by the formulae [3.1.21] and [3.1.22], which in the current context have the form

$$V \approx V_K^0 = -\frac{R_K^0}{R_K}, \quad V \approx V_K^\Delta = \frac{R_K^\Delta}{2R_K}, \tag{4.2.2}$$

$$\frac{R_K^\Delta}{2R_K} = \frac{\sum_{k=1}^K (\Delta\alpha_k)^2}{2\sum_{k=1}^K \alpha_k^2}, \tag{4.2.3}$$

$$R_K = -\frac{\sum_{k=1}^K \alpha_k \Delta\alpha_k}{\sum_{k=1}^K \alpha_k^2}, \tag{4.2.4}$$

and also

$$\sigma^2 \approx B_K^0 = \frac{1}{K}\sum_{t=0}^{K-1} (\Delta\mu_{K+1})^2. \tag{4.2.5}$$

Figures 4.5 and 4.6 illustrate the deviation of two estimates [4.2.2], calculated on the trajectories of SEs.

In particular, for $V = 0,8$, both estimates [4.2.2] are slightly differed in them (see Figure 4.5), but for $V = 1,95$, the *a priori* estimate V_K^Δ is less biased than the optimal estimate V_K^0 (see Figure 4.6).

Now, using numerically constructed DMD paths $\alpha(k)$, $k \geq 0$, we can estimate the basic parameters of DMD according to the statistics [4.2.2]–[4.2.4] and compare the obtained estimates with the real parameters of DMDs.

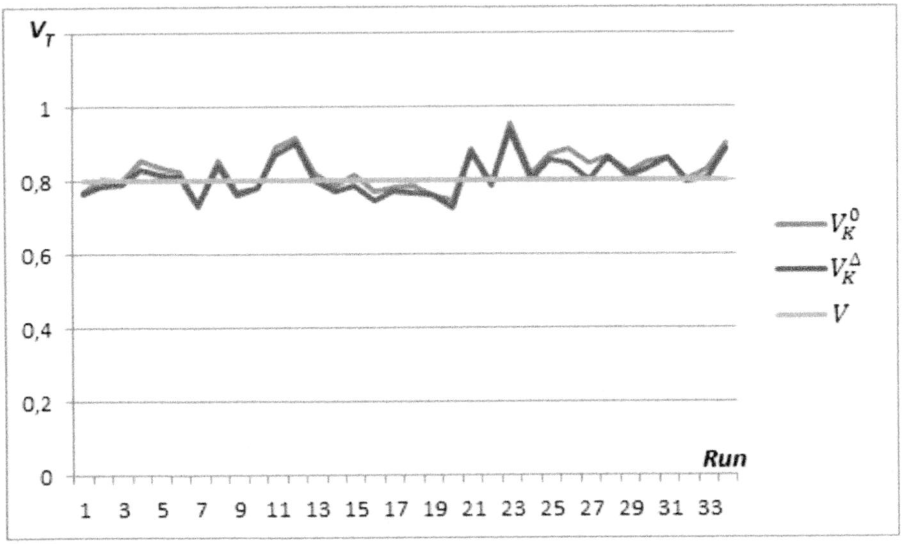

Figure 4.5. A priori *estimates* V_K^0, V_K^Δ *of drift parameter* $V = 0,8$ *for 33 launches of trajectory generation* $S_N(k)$, $N = 240$, $1 \leq k \leq 400$

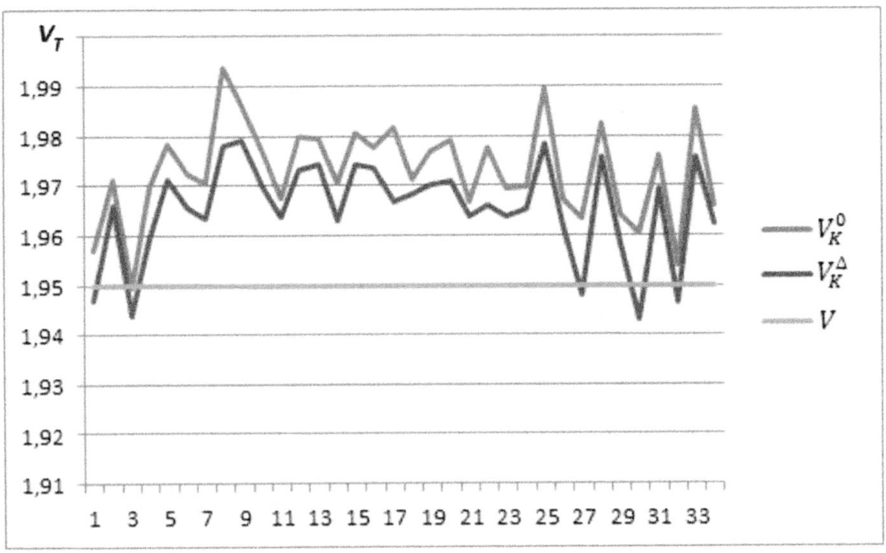

Figure 4.6. A priori *estimates* V_K^0, V_K^Δ *of drift parameter* $V = 1,95$ *for 33 launches of trajectory generation* $S_N(k)$, $N = 240$, $1 \leq k \leq 400$

We consider only three control calculations with simulation of trajectories according to the given parameters: the estimation of these parameters by DMD trajectories, according to formulae [4.2.2]–[4.2.4] and their comparison with the theoretical values.

Figure 4.7 shows the calculation table for numerical simulation of the DMD [4.1.1] trajectories for given parameters of the model $V = 0,5$ and $\sigma = 3$. At the same time, the covariance statistics, which form the basis for calculating the statistical estimates of the model parameters, are also calculated on DMD trajectories, by means of recurrent procedures.

	Unif.[0,1]	N(0,1)	α(k)	Δα(k+1)	α²(k)	Δα²(k+1)	α(k)Δα(k+1)	
	0,48692	-0,032793	-0,0328		0,0011	0	-0,18983903	σ₀²=Σα²(k)/N
	0,972836	1,9242092	5,75623	5,78902	33,134	33,5128	-56,5334853	11,77702315
E(B:B)	0,010323	-2,314384	-4,065	-9,8213	16,525	96,45729	-23,927584	
0,50061692	0,900527	1,2845581	1,82116	5,88619	3,3166	34,64726	-5,47101341	V=ΣΔα²/2Σα²
	0,242634	-0,697855	-1,183	-3,0041	1,3995	9,024872	4,216622888	0,501851668
E(C:C)	0,082981	-1,385294	-4,7474	-3,5644	22,538	12,70487	-12,8508711	
0,0003758	0,544225	0,1110849	-2,0404	2,70694	4,1634	7,327537	7,716381144	V=-ΣαΔα/Σα²
	0,054727	-1,600651	-5,8222	-3,7817	33,898	14,30154	-26,4879482	0,5018514
D(C:C)	0,707514	0,5461374	-1,2727	4,5495	1,6197	20,69793	-0,41126939	
0,98387447	0,458428	-0,104394	-0,9495	0,32315	0,9016	0,104429	2,082292466	W=1-V/2
	0,186934	-0,889252	-3,1425	-2,193	9,8754	4,809239	-13,5895718	0,749074166
	0,820618	0,9177217	1,18191	4,32442	1,3969	18,70064	0,25258224	
	0,605735	0,2682203	1,39561	0,21371	1,9477	0,045671	0,519440053	C=2V·W
	0,63933	0,3566672	1,76781	0,37219	3,1251	0,138529	4,632617133	0,751848239
	0,878627	1,1681487	4,38835	2,62054	19,258	6,867239	9,664700441	
	0,928609	1,4655097	6,5907	2,20235	43,437	4,850362	-9,47528039	σ²=C·σ₀²
	0,732116	0,6192263	5,15303	-1,4377	26,554	2,066905	4,449137421	8,85453412
	0,874235	1,1466392	6,01643	0,8634	36,197	0,745463	-30,180409	
	0,251629	-0,669371	1,0001	-5,0163	1,0002	25,16356	1,456389494	σ=√σ²
	0,742831	0,6520967	2,45634	1,45624	6,0336	2,12063	-2,07195797	2,975656922
	0,551012	0,1282192	1,61283	-0,8435	2,6012	0,711515	3,482261652	
	0,838547	0,9885055	3,77193	2,1591	14,227	4,661722	-14,416975	
	0,259333	-0,645403	-0,0502	-3,8222	0,0025	14,60901	-0,14377042	
	0,827787	0,9454578	2,81125	2,86149	7,9031	8,188153	-14,938653	
	0,09633	-1,302751	-2,5026	-5,3139	6,2631	28,23732	0,677963527	
	0,305935	-0,507405	-2,7735	-0,2709	7,6925	0,073387	-11,2584469	
	0,813489	0,8908288	1,28572	4,05925	1,6531	16,47751	-4,01996053	
	0,203858	-0,827919	-1,8409	-3,1266	3,3889	9,775735	-4,97075478	
	0,723492	0,5932454	0,85929	2,70018	0,7384	7,290992	2,261609262	
	0,846262	1,0205333	3,49124	2,63196	12,189	6,92719	12,25036832	
	0,96007	1,7515018	7,00013	3,50888	49,002	12,31226	-27,6197482	
	0,440968	-0,148514	3,05452	-3,9456	9,3301	15,56781	0,199257072	
	0,702232	0,5308313	3,11975	0,06523	9,7329	0,004255	-2,17458153	
	0,613179	0,2876137	2,42272	-0,697	5,8696	0,485859	5,905359087	
	0,888062	1,2162841	4,86021	2,43749	23,622	5,941372	12,07322449	

Parameters

	Theo	Run 1	Run2	Run 3	Run 4
σ	3	2,975656922	2,989673003	3,001539653	3,020546759
V_1	0,5	0,501851668	0,503952936	0,500753927	0,501445435
V_2	0,5	0,5018514	0,503917797	0,50075284	0,50141154
C_1	0,75	0,751848239	0,753937311	0,750753359	0,751443346
C_2	0,75	0,751847973	0,753902448	0,750752273	0,751409547

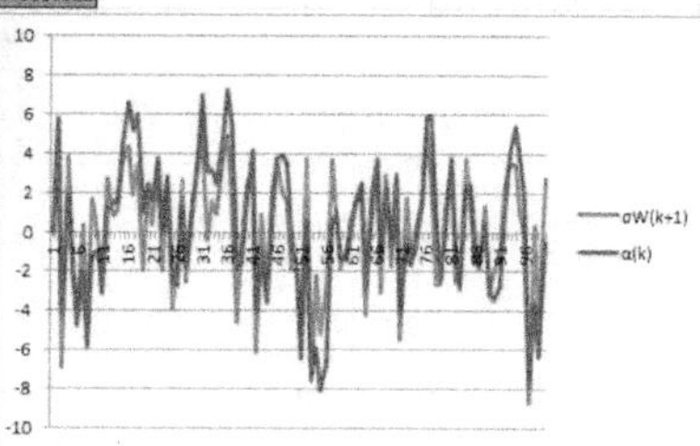

Figure 4.7. *Simulation of DMD trajectories with Gaussian stochastic component for $V = 0,5$ and $\sigma = 3$ and the corresponding parameter estimates*

The parameter estimation is shown in the right-hand side of the table and contains the estimates not only of the main parameters of the model V and σ, but also of additional parameters C and W (stationary coefficients).

We draw attention to the presence of two estimates V_1 and V_2 (as well as C_1 and C_2). This is due to the existence of two different statistical estimates for the same parameter V.

Such redundancy of estimates have an important role in verifying the model's adequacy for a number of sample values for a decision: to accept or reject a sample for the analysis based on the DMD model.

Figure 4.8 shows the numerical simulation of DMD trajectories for cardinally other values of model parameters $V = 1,9$ and $\sigma = 10$.

B:B	C:C					Δα(k+1)=-Vα(k)+σW(k+1)
Unif.[0,1]	N(0,1)	α(k)	Δα(k+1)	α²(k)	Δα²(k+1)	α(k)Δα(k+1)
0,77248	0,74704	0,74704		0,5581	0	-3,56691014
0,368611	-0,335535	-4,0277	-4,7747	16,222	22,79799	24,76104851
0,083788	-1,380031	-10,175	-6,1477	103,54	37,79438	-130,404947
0,25728	-0,651755	2,64031	12,8157	6,9713	164,2424	-33,866148
0,217402	-0,780997	-10,186	-12,827	103,76	164,5208	-178,315507
0,426678	-0,184838	7,31925	17,5055	53,571	306,4427	-339,466653
0,000582	-3,247341	-39,061	-46,38	1525,7	2151,103	-2735,74208
0,338075	-0,417722	30,9774	70,0382	959,6	4905,345	-902,496936
0,998522	2,9723118	1,84343	-29,134	3,3982	848,7905	-15,520487
0,311471	-0,491686	-6,5759	-8,4194	43,243	70,88581	-30,9728174
0,218158	-0,778428	-1,8659	4,71002	3,4817	22,18426	-9,47076109
0,560815	0,1530367	3,2097	5,07563	10,302	25,76202	-4,52666111
0,680398	0,4688128	1,7994	-1,4103	3,2378	1,988962	6,230943113
0,754326	0,6881649	5,26219	3,4628	27,691	11,99096	3,462174883
0,8567	1,06561	5,92013	0,65793	35,048	0,432877	62,49294467
0,985387	2,1804255	16,4761	10,556	271,46	111,4294	-627,306981
0,249235	-0,676899	-21,598	-38,074	466,45	1449,603	-1050,43784
0,776423	0,760169	27,0395	48,637	731,13	2365,555	-1259,62554
0,684039	0,4790241	-19,545	-46,585	382,02	2170,136	-533,86767
0,163011	-0,982159	7,76915	27,3144	60,36	746,0775	-133,778204
0,402928	-0,245776	-9,45	-17,219	89,303	296,4992	-262,952165
0,838194	0,9870624	18,3756	27,8256	337,66	774,2654	-599,249446
0,591054	0,230258	-14,235	-32,611	202,65	1063,484	-385,122717
0,500251	0,0006301	12,8182	27,0537	164,31	731,9036	-135,091571
0,916447	1,3815632	2,27922	-10,539	5,1948	111,0708	-3,70002274
0,606695	0,2707147	0,65585	-1,6234	0,4301	2,635337	14,5319326
0,990367	2,3403573	22,8133	22,1575	520,45	490,9531	-1269,315
0,109462	-1,229394	-32,826	-55,639	1077,5	3095,724	-2074,43605
0,532908	0,0825815	30,3691	63,1951	922,28	3993,616	-1650,82381
0,630915	0,3342779	-23,989	-54,359	575,49	2954,856	-1209,87157
0,686287	0,4853535	26,444	50,4335	699,29	2543,537	-1268,78496
0,589542	0,2263674	-21,536	-47,98	463,8	2302,08	-842,047392
0,427841	-0,181873	17,5636	39,0996	308,48	1528,778	-262,525517
0,96729	1,8423804	2,61653	-14,947	6,8462	223,416	3,672140874
0,738095	0,6374848	4,01997	1,40344	16,16	1,96964	-37,3948904

E(B:B)	E(C:C)	D(C:C)
0,50100369	0,00395742	1,00046054

$\sigma_0^2 = \Sigma\alpha^2(k)/N$	$V \approx \Sigma\Delta\alpha^2/2\Sigma\alpha^2$	$V \approx -\Sigma\alpha\Delta\alpha/\Sigma\alpha^2$	$W = 1 - V/2$	$C = 2V \cdot W$	$\sigma^2 = C \cdot \sigma_0^2$	$\sigma = \sqrt{\sigma^2}$
510,8190446	1,896739665	1,896733414	0,051630167	0,195857973	100,0479825	10,00239884

V= 1,9 σ= 10

Parameters	Theo	Run 1	Run2	Run 3	Run 4
σ	10	10,00239884	10,00716398	10,01932857	10,02815879
V_1	1,9	1,896739665	1,900180846	1,901707953	1,902593721
V_2	1,9	1,896733414	1,900177849	1,901684477	1,902580486
C_1	0,19	0,195857973	0,189674444	0,186922767	0,185324575
C_2	0,19	0,195869184	0,18967984	0,186965103	0,185348467

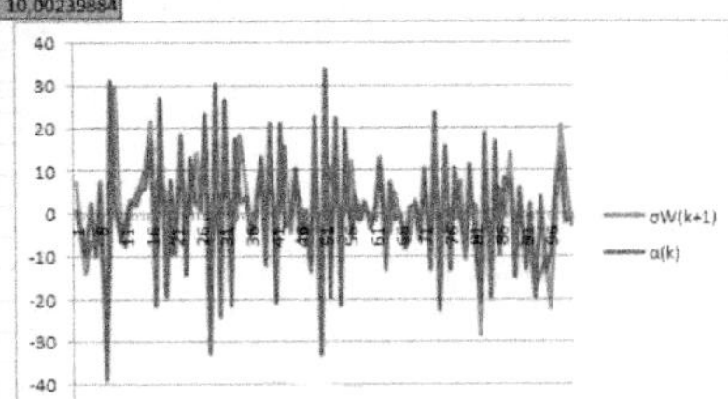

Figure 4.8. *Simulation of DMD trajectories with Gaussian stochastic component for $V = 1,9$ and $\sigma = 10$ and the corresponding parameter estimates*

Figure 4.9 represents the numerical simulation of DMD trajectories [4.2.1] with a martingale non-Gaussian stochastic component for $V = 1,9$ and $\sigma = 10$ and corresponding estimates of parameters. A non-Gaussian stochastic component is represented by a sequence of centered uniformly distributed random variables $U(-0,5, +0,5)$. The accuracy of such non-Gaussian DMD estimates are not significantly different from the accuracy of Gaussian DMD estimates.

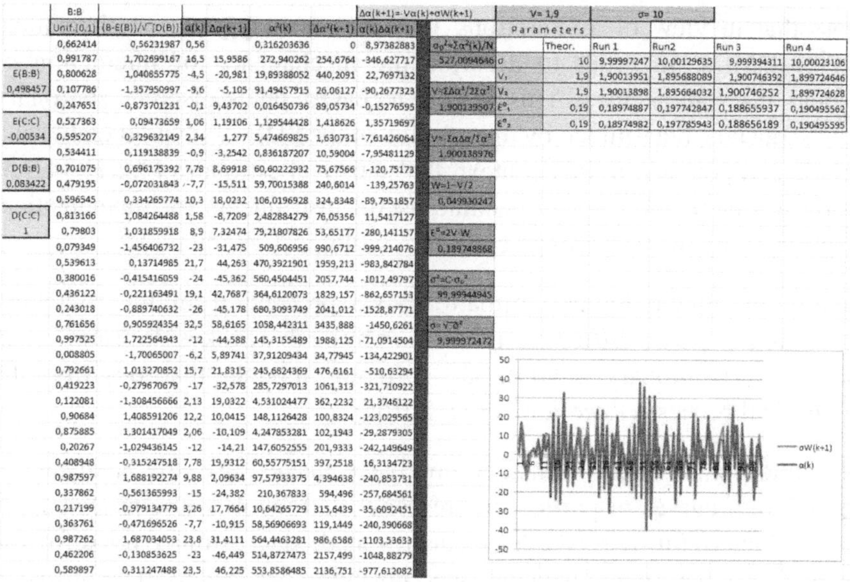

B:B Unif.[0,1]	(B-E(B))/√[D(B)]	α(k)	Δα(k+1)	α²(k)	Δα²(k+1)	α(k)Δα(k+1)
0,662414	0,56231987	0,56		0,316203636	0	8,97382883
0,991787	1,702699167	16,5	15,9586	272,940262	254,6764	-346,627717
0,800628	1,040855775	-4,5	-20,981	19,89388052	440,2091	22,7697132
0,107786	-1,357950997	-9,6	-5,105	91,49457915	26,06127	-90,2677323
0,247651	-0,873701231	-0,1	9,43702	0,016450736	89,05734	-0,15276595
0,527363	0,09473659	1,06	1,19106	1,129544428	1,418626	1,35719697
0,595207	0,329632149	2,34	1,277	5,474669825	1,630731	-7,61426064
0,534411	0,119138839	-0,9	-3,2542	0,836187207	10,59004	-7,95481129
0,701075	0,696175392	7,78	8,69918	60,60222932	75,67566	-120,75173
0,479195	-0,072031843	-7,7	-15,511	59,70015388	240,6014	-139,25763
0,596545	0,334265774	10,3	18,0232	106,0196928	324,8348	-89,7951857
0,813166	1,084264488	1,58	-8,7209	2,482884279	76,05356	11,5417127
0,79803	1,031859918	8,9	7,32474	79,21807826	53,65177	-280,141157
0,079349	-1,456406732	-23	-31,475	509,606956	990,6712	-999,214076
0,539613	0,13714985	21,7	44,263	470,3921901	1959,213	-983,842784
0,380016	-0,415416059	-24	-45,362	560,4504451	2057,744	-1012,49797
0,436122	-0,221163491	19,1	42,7687	364,6120073	1829,157	-862,657153
0,243018	-0,889740632	-26	-45,178	680,3093749	2041,012	-1528,87771
0,761656	0,905924354	32,5	58,6165	1058,442311	3435,888	-1450,6261
0,997525	1,722564943	-12	-44,588	145,3155489	1988,125	-71,0914504
0,008805	-1,70065007	-6,2	5,89741	37,91209434	34,77945	-134,422901
0,792661	1,013270852	15,7	21,8315	245,6824369	476,6161	-510,63294
0,419223	-0,279670679	-17	-32,578	285,7297013	1061,313	-321,710922
0,122081	-1,308456666	2,13	19,0322	4,531024477	362,2232	21,3746122
0,90684	1,408591206	12,2	10,0415	148,1126428	100,8324	-123,029565
0,875885	1,301417049	2,06	-10,109	4,247853281	102,1943	-29,2879305
0,20267	-1,029436145	-12	-14,21	147,6052555	201,9333	-242,149649
0,408948	-0,315247518	7,78	19,9312	60,55775151	397,2518	16,3134723
0,987597	1,688192274	9,88	2,09634	97,57933375	4,394638	-240,853731
0,337862	-0,561365993	-15	-24,382	210,367833	594,496	-257,684561
0,217199	-0,979134779	3,26	17,7664	10,64265729	315,6439	-35,6092451
0,363761	-0,471696526	-7,7	-10,915	58,56906693	119,1449	-240,390668
0,987262	1,687034053	23,8	31,4111	564,4463281	986,6586	-1103,53633
0,462206	-0,130853625	-23	-46,449	514,8727473	2157,499	-1048,88279
0,589897	0,311247488	23,5	46,225	553,8586485	2136,751	-977,612082

E(B:B) 0,498457

E(C:C) -0,00534

D(B:B) 0,083422

D(C:C) 1

Δα(k+1)=-Vα(k)+σW(k+1)

α0²=Σα²(k)/N 527,0094646

V=ΣΔα²/2Σα² 1,900139507

V=-ΣαΔα/Σα² 1,900138976

W=1−V/2 0,049930247

ε°=2V·W 0,189748868

σ²=C·σ0² 99,99944945

σ=√σ² 9,999972472

V= 1,9		σ= 10			
Parameters	Theor.	Run 1	Run2	Run 3	Run 4
σ	10	9,99997247	10,00129635	9,999394311	10,00023106
V₁	1,9	1,90013951	1,895688089	1,900746392	1,899724646
V₂	1,9	1,90013898	1,895664032	1,900746252	1,899724628
ε°₁	0,19	0,18974887	0,197742847	0,188655937	0,190495562
ε°₂	0,19	0,18974982	0,197785943	0,188656189	0,190495595

Figure 4.9. *Simulation of DMD trajectories with the Gaussian stochastic component for* $V = 1,9$ *and* $\sigma = 10$ *and the corresponding parameter estimates*

4.3. DMD and modeling of the dynamics of macromolecules in biophysics

In biological processes with equilibrium, the dynamics of concentration, or frequencies, of a predefined characteristic, can be described by a mathematical model of binary SEs, based on statistical data of elementary hypothesis validation about the presence or absence of a predefined attribute in the set of elements that make up a complex system.

It is assumed that:

1) All the elements that make up the system can gain or lose the attribute over time, that is, the frequency of the attribute is a dynamic variable.

2) The basic objective of our study are the SEs, characterized by relative frequencies of the presence or absence of the attribute in a sample of fixed volume, at each time instant.

3) It is assumed that the result of the next experiment, at time instant $k + 1$, depends on the average result of the present experiment, at time instant k. This relationship is concretized in Chapters 1–3 as *basic assumptions*.

Note that in view of assumptions 1–3, we can apply the mathematical model of DMDs, in order to describe the mechanisms of interaction of biological macromolecules. This model is based on a discrete stationary random sequence, determined by the solution of the SDEs with predictable and stochastic components. In this context, the statistical estimates [4.2.2]–[4.2.4] of DMD parameters V and σ^2 become useful instruments for mathematical description of the mechanisms of collective biological interactions.

In the subsequent logical formations, we proceed from the following consideration: two diffusion models, physical (Stokes–Einstein) and mathematical (stationary DMD), are different models that describe the same physical process (Koroliouk *et al.* 2016a).

The traditional Stokes–Einstein approach makes use the analysis of autocorrelation curves of particle fluorescence and involves the procedure of exponential curve fitting by the least-squares method to estimate the diffusion coefficient of experimental data and accompanying parameters. The diffusion coefficient D derives from the characteristic decay time of the correlation function (Schwille *et al.* 1997). This method gives, at first glance, good evaluation results, but in reality the values of the parameters generally may not reflect the actual properties of the sample or have no physical meaning.

The proposed method of statistical analysis of real measurements is specially designed for the analysis of microfluorescence measurements in the area where the observation of physical data (biological, chemical, etc.) may have no logical reason in the form of a simple physical theory.

4.3.1. *The model motivation*

The basis of obtaining the statistical data is the method of dynamic monitoring of macromolecule FCS. It calculates the number of the fluctuations of fluorescence-labeled molecules in a small (confocal) range of observations.

The main objective of the analysis is the fluctuation of fluorescence intensity α_t, that is, the difference between its current value at the time instant $t \geq 0$ and its average value. The intensity of the fluorescence is proportional to the number of labeled molecules observed at the instant t. The molecules can freely diffuse: to enter and go beyond the observation volume, or undergo chemical reactions. Thus, as a result, there are combinations of motions of different types, characterized by different drift and diffusion parameters.

The developed mathematical model of stationary Gaussian diffusion is determined by a continuous process such as the Ornstein–Uhlenbeck process. It is motivated by an important particular case of the motion of particles in a viscous liquid medium, known in physics as a trajectory of solution of the Langevin equation.

Consider the equation:

$$m\dot{v}(t) = -\beta v(t) + \mu_t, \quad t \geq 0, \qquad [4.3.1]$$

where M is the particle mass, $v(t)$ the particle velocity, β the average viscosity and μ_t a noise factor, which will be further refined.

Here, the component of the "force" $m\dot{v}(t)$ slows the particle in proportion to the "velocity" $v(t)$, and the presence of "noise" is associated with chaotic particle collisions with molecules of the medium (due to the thermal motion of the latter).

This formulation uses, as "noise," a "random process" $\mu_t = \dot{W}_t$, where $W = \{W_t\,,\ t \geq 0\}$ is a stochastic process of Brownian motion.

In order to give meaning to the equation

$$m\dot{v}(t) = -\beta v(t) + \dot{W}_t, \quad t \geq 0, \qquad [4.3.2]$$

known in physics as the Langevin equation, should give the probabilistic content of the derivative of the process $\dot{W}_t$, $t \geq 0$. This is done using the concept of Itô stochastic integral.

4.3.2. *Statistical model of a stationary DMD*

As was noted in section 4.3, the DMD model is considered as a stationary Gaussian sequence α_t, $t \in N_0 = \{0, 1, 2, ...\}$, which is given by the solution of the SDE (Ch. 3, [3.1.1]).

A wide-sense stationary, random sequence α_t, $t = 0, 1, 2, ...$, is characterized by two numerical parameters:

V_0 : regression parameter of the predicted component $(0 < V < 2)$;

σ^2: square variance of the stochastic component $\sigma\Delta W(t)$, $t = 0, 1, 2, ...$, is determined by a sequence of independent equally distributed Gaussian random variables with parameters $(0,\ \sigma^2)$.

A priori statistics are used to estimate the numerical parameters V_0 (drift) and σ^2 (square variance of the stochastic component) of a model generated by the solutions of the SDE [4.1.13].

The stationarity of DMD allows for the use of covariations of two-component vector $(\alpha_t,\ \Delta\alpha_t)$, $t \geq 0$, which generates the covariance statistics (see equation [3.1.24]):

$$R_T := \frac{1}{T}\sum_{t=0}^{T-1}\alpha_t^2,\quad R_T^0 = \frac{1}{T}\sum_{t=0}^{T-1}\alpha_t\Delta\alpha_t,\quad R_T^\Delta := \frac{1}{T}\sum_{t=0}^{T-1}(\Delta\alpha_t)^2. \qquad [4.3.3]$$

The statistics [4.3.3] generate two *a priori* estimates of the parameter V_0 of the drift component (see [4.2.2]):

$$V_0 \approx V_T^0 = -\frac{R_T^0}{R_T},\quad V_0 \approx V_T^\Delta = \frac{R_T^\Delta}{2R_T}, \qquad [4.3.4]$$

which are strongly consistent and asymptotically unbiased.

Naturally, the following obvious estimate takes place:

$$\sigma_0^2 \approx \sigma_{0T}^2 = R_T. \qquad [4.3.5]$$

The stationarity condition [3.1.4] generates the stochastic component variance σ^2:

$$\sigma^2 \approx \sigma_T^2 = \mathcal{E}_T^0 R_T, \qquad [4.3.6]$$

where the coefficient of stationarity is

$$\mathcal{E}_T^0 := V_T^0(2 - V_T^0). \qquad [4.3.7]$$

REMARK 4.3.1.– Formula [4.3.7], by condition $0 < V_0 < 1$, can also be used as control of the dynamics:

$$\ln \operatorname{cov}(\alpha_t,\ \alpha_{t+s}) = \ln \sigma_0^2 + s \ln(1 - V_0),\quad s, t \geq 0. \qquad [4.3.8]$$

Thus, on a logarithmic scale, the covariance of stationary DMD is characterized by a linear function $a - sb$, where, by definition,

$$a := \ln \sigma_0^2,\quad b := -\ln(1 - V_0). \qquad [4.3.9]$$

The verification of DMD stationarity can be done using the following parameter estimates a and b:

$$a \approx a_T = \ln R_T,\quad b \approx b_T = -\ln(1 - V_T). \qquad [4.3.10]$$

4.3.3. *Stokes–Einstein kinetic diffusion model*

A three-dimensional Brownian motion of n particles which move in a cube with the side L, centered at the origin of Cartesian coordinates, is obtained by numerical modeling (see also Fischer 2008).

The Cartesian coordinates of each particle were modeled as independent discrete scalar Brownian motions. Let W_t denotes the value of a scalar Brownian motion at the instant $t = 0, 1, 2, \ldots$. The following scheme of recurrent numerical simulation is used:

$$W_t = W_{t-1} + \Delta W_t, \quad t = 0, 1, 2, \ldots, \qquad [4.3.11]$$

where every ΔW_t is a sequence of independent random variable of the form $\sqrt{2D\delta}N(0,1)$. Here $N(0,1)$ means a normal standard random variable, δ the sampling time interval and D the diffusion coefficient, given by

$$D = \frac{kT}{6\pi r \eta}, \qquad [4.3.12]$$

where kT is the kinetic energy, r the particle radius and η the medium viscosity.

The periodic boundary conditions were the kinetic energy at the boundary of a elementary cube. The initial position of each particle was taken as a three-dimensional random variable, uniformly distributed inside the elementary cube. As the region of interest (ROI), we consider a spherical measurement area with the center at the origin of coordinates and diameter D_R. At every time instant t, the number α_t of the particles inside the ROI were registered with the corresponding coordinates.

Two cycle simulations were carried out. First, all the particles had a given same diffusion coefficient D.

Then a mixture of two families of particles with different diffusion coefficients D_1 and D_2 were summed up. The volume fraction of slowly diffusing particles was denoted f.

The values of the parameters adopted in the simulation were the diameter ROI $1\,\mu\text{m}$; 10 particles expected in ROI, giving the density 19.099 particles/μm^3. Assuming the sampling frequency $50\,\text{kHz}$ (frequency of measurements: $\delta = 20\,\mu\text{s}$) over time simulation $10\,\text{s}$ (500K samples).

The first simulation run was performed using diffusion coefficient at different values, respectively: 10, 30, 50, 80, 100 $\mu m^2/s$. In the second run, two sets of particles together are present in the ROI, namely 10 and 100 $\mu m^2/s$. By changing the number of particles ratio between these two sets from 0 to 1, we obtain the basic parameters of the samples, which are given in

Parameter	Description	Value/Range	Unit
D_R	ROI diameter	1	μm
L	Side of the cubical simulation box (periodic boundary conditions)	10	μm
N	Total number of particles	19.099	$\sharp$
n_{ROI}	Average number of expected particles in ROI	10	$\sharp$
P	Average density of particles	19.099	$\sharp/\mu m^3$
δ	Sampling time interval	20	μs
f_S	Sampling rate	50	kHz
T	Total simulation time	10	S
S	Total number of samples	500,000	Samples
D	Diffusion coefficient (first cycle of simulation)	10, 30, 50, 80, 100	$\mu m^2/s$
D_1, D_2	Diffusion coefficient (second cycle of simulation)	10, 100	$\mu m^2/s$
F	Volume fraction of slowly diffusing particles (second cycle of simulation)	0, 0.1, 0.3, 0.5, 0.7, 0.9, 1	

Table 4.1. *The numerical parameters for samples simulation using the Stokes–Einstein kinetic diffusion model*

4.3.4. *Verification of the model of stationary DMD by direct numerical simulation*

The verification of this model by means of numerical simulation was carried out using numerical simulation of a stationary DMD trajectories (4.3.1) and (4.3.2) for given parameters V and σ^2.

Next, according to [4.3.11] and [4.3.12], the values of estimates V_T^0, σ_{0T}^2 and σ_T^2 are calculated in comparison with the initial (theoretical) values of parameters V and σ.

For different series of random numbers generation (run), we have the following result for the direct numerical simulation.

Below, there exposed a model of numerical simulation and parameter estimation, in two cases $V = 0,2; \sigma = 10$ and $V = 0,8; \sigma = 10$ (Figures 4.10 and 4.11).

In the calculations, we have used the sampling lengths of 30,000 sample values.

The model verification was carried out for the whole range of values V $(0 < V < 2)$ and σ $(\sigma > 0)$, that is, a large series of calculations for many combinations of theoretical parameters V and σ^2.

The numerical simulation of DMD trajectories, using *a priori* statistics [4.3.4] and [4.3.5], makes it possible to check the adequacy of statistical estimates of parameters V and σ^2.

4.3.5. *Numerical verification of DMD characteristics using the model of Stokes–Einstein*

The data based on the Stokes–Einstein diffusion model represent the dynamics of the number of numerically simulated particles calculated in a limited amount in preset time intervals (20 μs), for a full-time observation interval of about 0.328 s (16,385 sample values). The simulation generates a mixture of two Brownian motions: slow ($D = 10$) and fast ($D = 100$). The selected data are obtained for various proportions of mixing of fast and slow particles with the following proportions:$(0|1), (0,1|0,9), (0,3|0,7), (0,5|0,5)$, $(0,7|0,3), (0,9|0,1)$ and $(1|0)$. For each value of the mixing proportions, several samples are generated based on different series of random number generation.

Now, seven such groups of the sample data are considered as trajectories of stationary DMD, determined by formulae [4.3.2] and [4.3.3] with unknown parameters V and σ, estimated by the statistics [4.3.8]–[4.3.12].

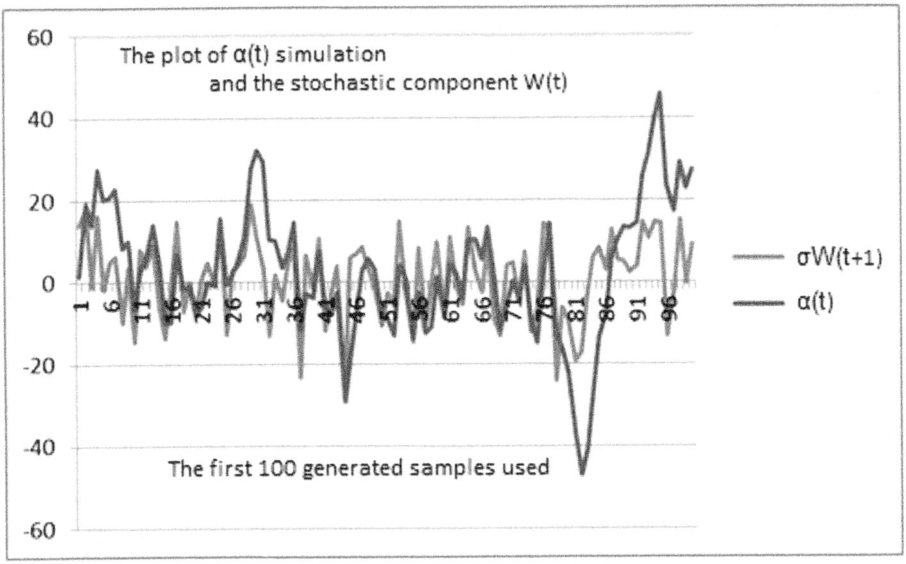

(a)

Δα(t+1)=-Vα(t)+σW(t+1)		V=0,2		σ=10	
PARAMETERS					
	Theor	Experimental evaluation			
		Run 1	Run2	Run 3	Run 4
σ	10	9,9878	9,978	10,057	10,0197
V	0,2	0,2066	0,201	0,1997	0,1989

(b)

Figure 4.10. *The case ($V = 0,2; \sigma = 10$): (a) a graph of numerical modeling of DMD $\alpha(t)$ and its stochastic components $\sigma W(t+1)$; (b) evaluation of the parameters of the simulated DMD for four different series of random number generation*

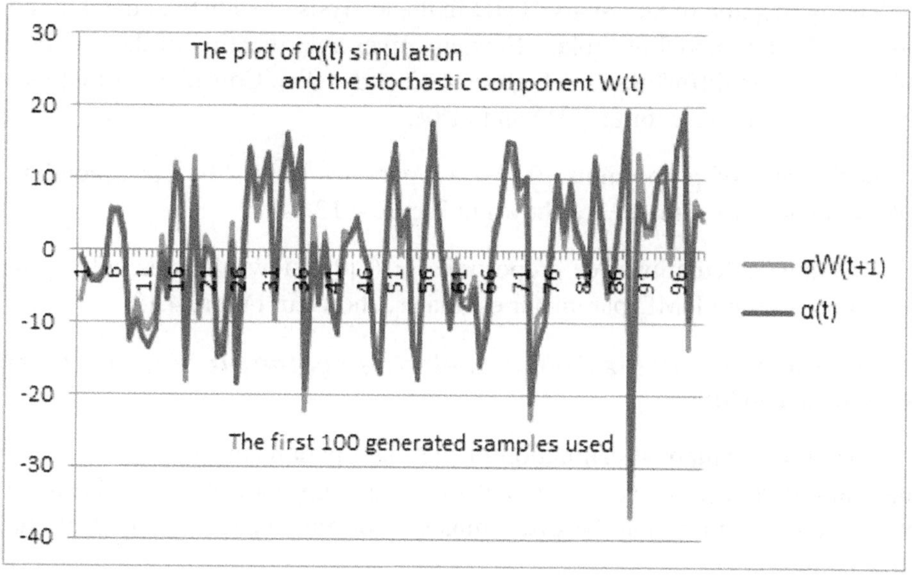

(a)

Δα(t+1)=-Vα(t)+σW(t+1)			V=0,8	σ=10	
PARAMETERS					
	Theor	Experimental evaluation			
		Run 1	Run2	Run 3	Run 4
σ	10	10,0237	9,951	10,012	10,0248
V	0,8	0,80173	0,792	0,8103	0,7977

(b)

Figure 4.11. *The case* $(V = 0,8; \sigma = 10)$*: (a) a graph of numerical modeling of DMD* $\alpha(t)$ *and its stochastic components* $\sigma W(t+1)$*; (b) evaluation of the parameters of the simulated DMD for four different series of random number generation*

The application of stationary DMD in the analysis of statistical data, based on the diffusion model of Stokes–Einstein, gives a verification of the mix rate of slow and fast diffusions (see Figures 4.12 and 4.13). Consider the first two extreme cases of a mixture: $(0|1)$ and $(1|0)$.

In the case of proportions $(0|1)$ (fast particles only), the corresponding DMD parameter estimates are shown in Figure 4.12.

The same calculations for proportions $= (1|0)$ (slow particles only) give the corresponding DMD parameter estimates, shown in Figure 4.13.

4.3.6. *The ability of the DMD model to detect the proportion of fast and slow particles*

The above computed estimation of the main parameters V and σ for fast and slow particles are very different in value. This indicates that the main parameters V and σ may be discriminators for any aspect ratio of fast and slow particles.

For verification, it is necessary to repeat the same procedure for calculating the model parameters for the proportions $(0,1|0,9), (0,3|0,7), (0,5|0,5),$ $(0,7|0,3), (0,9|0,1)$ and then compare the dynamics of the parameters V, σ and σ_0^2 according to the proportion of mixture of two Brownian movements: slow $(D = 10)$ and fast $(D = 100)$.

The trajectories obtained with algorithms of Stokes–Einstein kinetic diffusion model were calculated for seven values of the proportion of mixture $(0,1|0,9), (0,3|0,7), (0,5|0,5), (0,7|0,3)$ and $(0,9|0,1)$. Each simulation loop contains 500,000 sampling values that are used to evaluate the parameters V, σ and σ_0^2 by directly applying estimates [4.3.8]–[4.3.12].

The following results were obtained for the parameter σ.

The algorithm for estimating parameter V for different proportions of the mixture gives the following result.

Finally, for parameter σ_0^2, the following result holds.

A comparative analysis of the tables and relative graphs shows that the parameters σ and V have almost linear ability to distinguish the proportion of mixing of two ensembles of particles with different diffusion coefficients according to the diffusion Stokes–Einstein model (Figures 4.14 and 4.15).

However, parameter σ_0^2 has no such property (Figure 4.16).

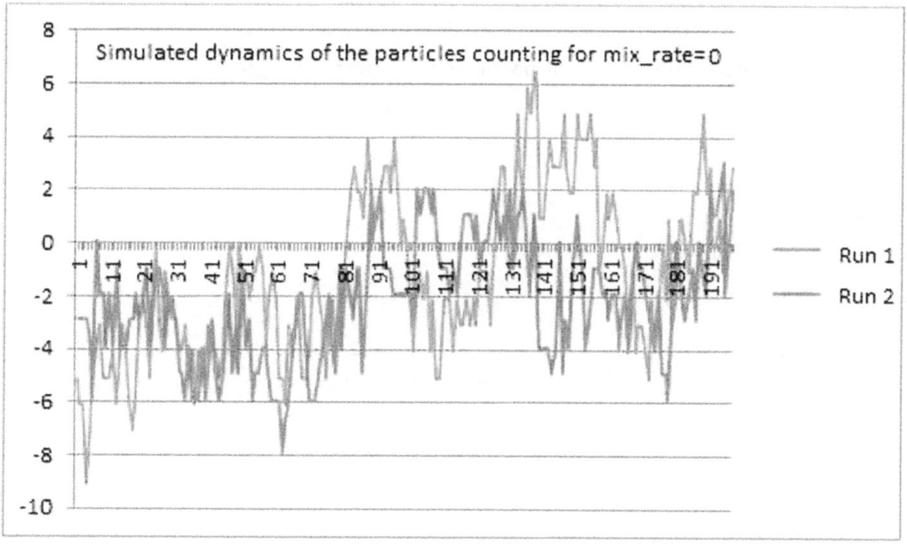

(a)

Parameter	Run 1	Run 2
$\sigma_0^2 \approx \Sigma\alpha^2(k)/T$	9,957267597	9,838731687
$V \approx \Sigma\Delta\alpha^2/2\Sigma\alpha^2$	0,151887047	0,151482925
$V \approx -\Sigma\alpha\Delta\alpha/\Sigma\alpha^2$	0,151887416	0,15148301
$W=1-V/2$	0,924056476	0,924258538
$C=2V \cdot W$	0,280704419	0,280018773
$\sigma^2=C \cdot \sigma_0^2$	2,795049017	2,755029577
$\sigma=\sqrt{\sigma^2}$	1,671840009	1,659828177

(b)

Figure 4.12. *Proportion of the mixture* = $(0|1)$*: (a) the trajectory of Stokes–Einstein kinetic model; (b) evaluation of the DMD stochastic model*

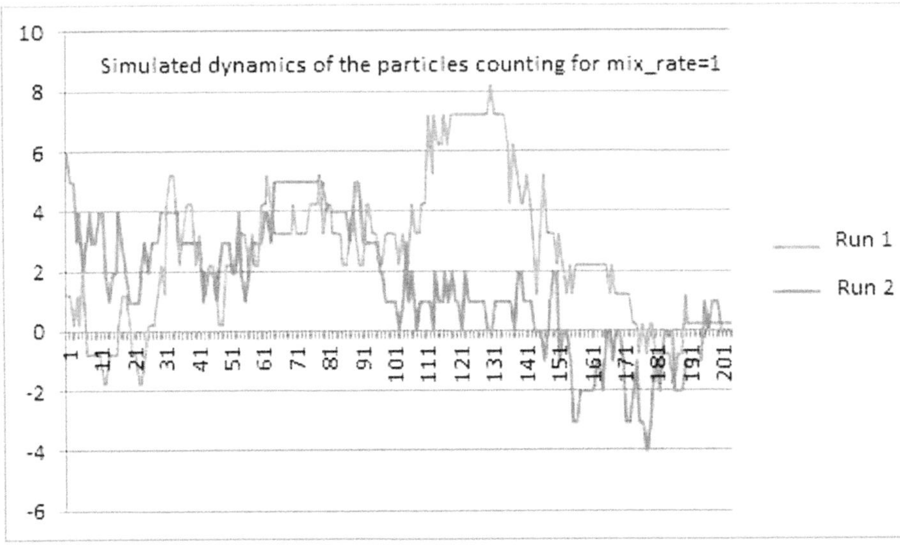

(a)

Parameter	Run 1	Run 2
$\sigma_0^2 \approx \Sigma\alpha^2(k)/T$	9,5117969	9,681002702
$V \approx \Sigma\Delta\alpha^2/2\Sigma\alpha^2$	0,049345364	0,049946949
$V \approx -\Sigma\alpha\Delta\alpha/\Sigma\alpha^2$	0,049345429	0,049947382
$W=1-V/2$	0,975327318	0,975026526
$C=2V\cdot W$	0,096255763	0,0973992
$\sigma^2=C\cdot\sigma_0^2$	0,91556527	0,942921915
$\sigma=\sqrt{\sigma^2}$	0,956851749	0,971041665

(b)

Figure 4.13. *Proportion of the mixture* = $(1|0)$*: (a) the trajectory of Stokes–Einstein kinetic model; (b) evaluation of the DMD stochastic model*

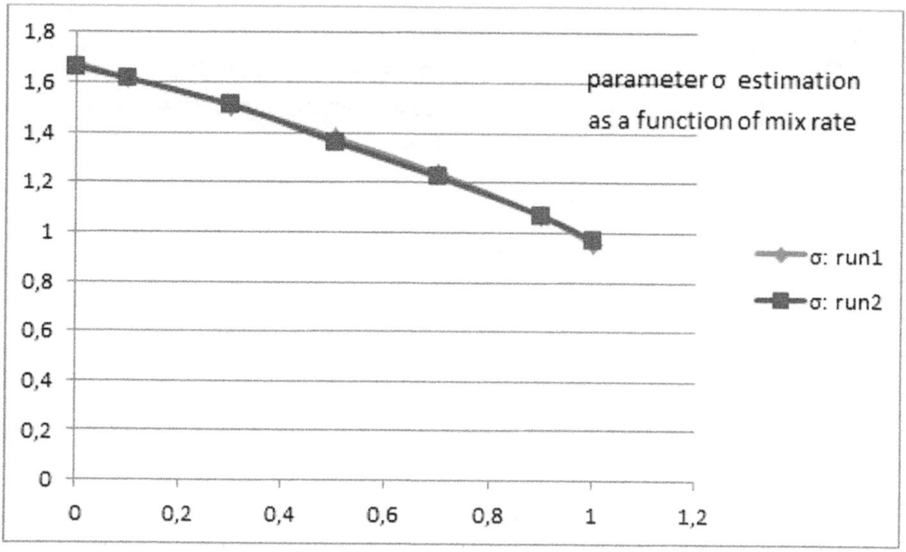

(a)

mix_rate	σ: run1	σ: run2
0	1,671841171	1,65982983
0,1	1,614770647	1,61374915
0,3	1,505486209	1,50890114
0,5	1,379160156	1,36214922
0,7	1,237259001	1,2220825
0,9	1,066198549	1,06760628
1	0,956851749	0,97104166

(b)

Figure 4.14. *Discriminating property of the parameter σ relative to mix ratio: (a) a graph of the estimated values of the parameter σ as a function of the proportion of the mixture; (b) parameter estimations σ for seven proportions obtained by two different random number generator runs*

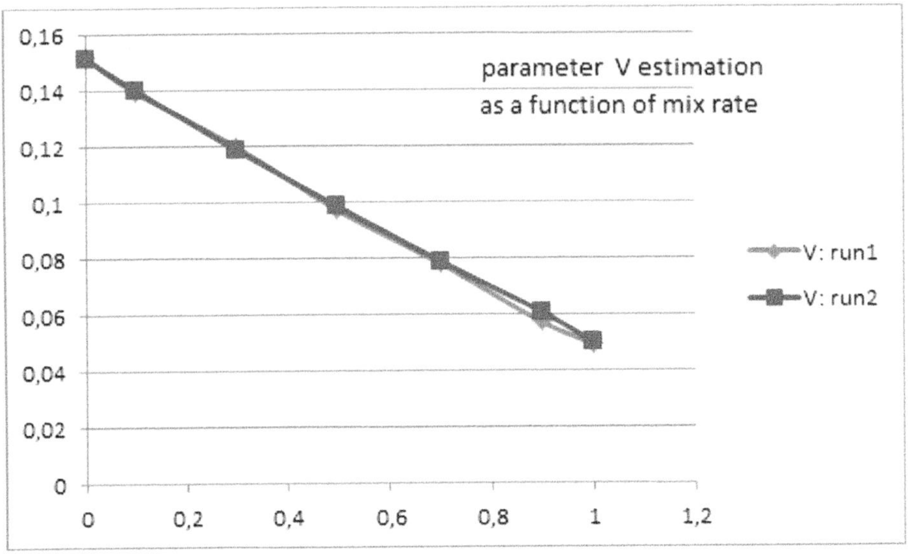

(a)

mix_rate	V: run1	V: run2
0	0,151888176	0,15148295
0,1	0,139527392	0,14036579
0,3	0,11998634	0,11873494
0,5	0,097219474	0,09873162
0,7	0,078169653	0,07911793
0,9	0,057210931	0,06086607
1	0,049345364	0,04994695

(b)

Figure 4.15. *Discriminating property of the parameter V relative to mix ratio: (a) a graph of the estimated values of the parameter V as a function of the proportion of the mixture; (b) parameter estimations V for seven proportions obtained by two different random number generator runs*

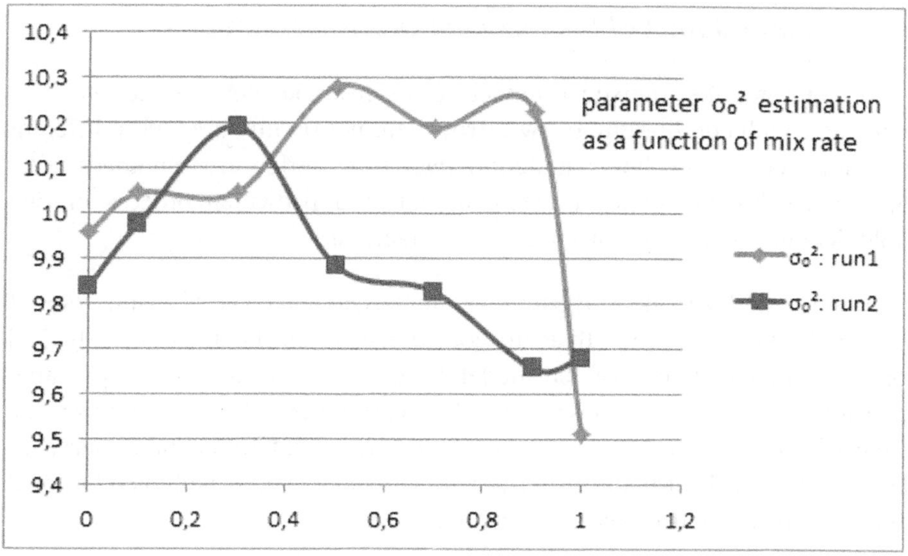

(a)

mix_rate	σ_0^2: run1	σ_0^2: run2
0	9,957213516	9,8387497
0,1	10,04474549	9,97661571
0,3	10,04756329	10,1927893
0,5	10,28223292	9,88438591
0,7	10,18985516	9,82710111
0,9	10,22754759	9,65693251
1	9,5117969	9,6810027

(b)

Figure 4.16. *Discriminating property of the parameter σ_0^2 relative to the mix ratio: (a) a graph of the estimated values of the parameter σ_0^2 as a function of the proportion of the mixture; (b) parameter estimations σ_0^2 for seven proportions obtained by two different random number generator runs*

4.3.7. *Interpretation of the mixes of Brownian motions*

To interpret the statistical model of macromolecules interaction as a dynamics, with equilibrium, of two Brownian motion mixtures with different diffusion coefficients, the simulation data time series are obtained using the discrete kinetic Stokes–Einstein model as a Brownian process of two biological molecules with different molar fractions.

Figures 4.17 and 4.18 illustrate a numerical analysis of a mixture of kinetic Brownian motions that are considered as DMD trajectories [4.2.1] for *unknown* parameters of the model V and σ. Accordingly, the prepared trajectories of statistical "observations" (namely, the values centered on the value of the equilibrium), according to the recurrent procedures, serve as data for statistical estimates of basic parameters (V, σ), as well as additional characteristics C and W (coefficients of stationarity).

We draw attention at two homogeneous series of calculations in two-colored tables, which means the estimation of parameters for two different numerical data samples corresponding to two different runs of modeling kinetic Brownian mixtures with independent generations of random numbers.

Figures 4.18 and 4.19 show the calculation of two components of kinetic Brownian motions: the first corresponds to the "pure fast" Brownian motion, and the second corresponds to the "pure slow" Brownian motion.

The "extreme cases" considered in Figures 4.18 and 4.19 show significant differences in the values. This points to the hypothesis that the basic parameters V and σ can be discriminant for any value of the proportion of fast and slow particles.

Note that the closeness of the values of the estimates V_1 and V_2 of the same parameter V in spreadsheets (Figures 4.19 and 4.20) shows the adequacy of the DMD model with a numerical set of sample values of the mixes of Brownian motions and the acceptability of these data for processing according to our algorithms. The next calculations of basic parameter estimates V and σ for the mixture proportions $(0,1|0,9)$, $(0,3|0,7)$, $(0,5|0,5)$, $(0,7|0,3)$, $(0,9|0,1)$ can establish the correspondence of parameters V, σ and σ_0^2 to the proportion of mixture of two Brownian movements: slow ($D = 10$) and fast ($D = 100$).

Centered_BM_DROI_1_npROI_10_fbound_0.0_Dfree_100_Dbound_100_fs_50000_T_10_LDROI_10_periodic_1

Time[s]	Run 1	Run 2		Rn1_Cnt	Rn2_Cnt										
	α	β		α(k)	β(k)	Δα(k+1)	Δα²(k+1)	α(k)Δα(k+1)	α²(k)	σ0²=Σα²(k)/N	Δβ(k+1)	Δβ²(k+1)	β(k)Δβ(k+1)	β²(k)	σ0²=Σβ²(k)/N
0	12	9		1,8828	-0,6392	0	0	0	3,54	9,814001497	0	0	0	0,41	8,820162232
0,00002	12	9	Eα	1,8828	-0,6392	3	9	5,64839458	3,54		4	16	-2,556953	0,41	
0,00004	15	13	10,11720181	4,8828	3,36076	0	0	0	23,8	V=ΣΔα²/2Σα²	-1	1	-3,360762	11,3	V=ΣΔβ²/2Σβ²
0,00006	15	12		4,8828	2,36076	-4	16	-19,5311928	23,8	0,153985522	1	1	2,3607618	5,57	0,162771217
0,00008	11	13	Eβ	0,8828	3,36076	1	1	0,88279819	0,78		-3	9	-10,08229	11,3	
0,0001	12	10	9,639238188	1,8828	0,36076	0	0	0	3,54	V=-ΣαΔα/Σα²	1	1	0,3607618	0,13	V=-ΣβΔβ/Σβ²
0,00012	12	11		1,8828	1,36076	3	9	5,64839458	3,54	0,153996547	1	1	1,3607618	1,85	0,162772631
0,00014	15	12		4,8828	2,36076	0	0	0	23,8		-1	1	-2,360762	5,57	
0,00016	15	11		4,8828	1,36076	3	9	14,6483946	23,8	W=1-V/2	-1	1	-1,360762	1,85	W=1-V/2
0,00018	18	10		7,8828	0,36076	0	0	0	62,1	0,923007239	4	16	1,4430472	0,13	0,918614392
0,0002	18	14		7,8828	4,36076	-3	9	-23,6483946	62,1		-2	4	-8,721524	19	
0,00022	15	12		4,8828	2,36076	4	16	19,5311928	23,8	C=2V-W	1	1	2,3607618	5,57	C=2V-W
0,00024	19	13		8,8828	3,36076	-5	25	-44,413991	78,9	0,284259503	-1	1	-3,360762	11,3	0,299047965
0,00026	14	12		3,8828	2,36076	0	0	0	15,1		0	0	0	5,57	
0,00028	14	12		3,8828	2,36076	3	9	11,6483946	15,1	σ²=C-σ0²	-2	4	-4,721524	5,57	σ²=C-σ0²
0,0003	17	10		6,8828	0,36076	-2	4	-13,7655964	47,4	2,789723186	1	1	0,3607618	0,13	2,637851565
0,00032	15	11		4,8828	1,36076	3	9	14,6483946	23,8		0	0	0	1,85	
0,00034	18	11		7,8828	1,36076	-2	4	-15,7655964	62,1	σ=√σ²	-2	4	-2,721524	1,85	σ=√σ²
0,00036	16	9		5,8828	-0,6392	2	4	11,7655964	34,6	1,670246445	2	4	-1,278476	0,41	1,624084839
0,00038	18	11		7,8828	1,36076	0	0	0	62,1		1	1	1,3607618	1,85	
0,0004	18	12		7,8828	2,36076	-2	4	-15,7655964	62,1		-1	1	-2,360762	5,57	
0,00042									,6		-2	4	-2,721524	1,85	
0,00044									,6		2	4	-1,278476	0,41	
0,00046									54		1	1	1,3607618	1,85	
0,00048									31		-3	9	-7,082285	5,57	
0,0005									78		3	9	-1,917715	0,41	
0,00052									31		-1	1	-2,360762	5,57	
0,00054									,8		-1	1	-1,360762	1,85	
0,00056									78		3	9	1,0822854	0,13	
0,00058									54		-4	16	-13,44305	11,3	
0,0006									54		7	49	-4,474667	0,41	
0,00062									54		-3	9	-19,08229	40,5	
0,00064									,6		2	4	6,7215236	11,3	
0,00066									,8		-1	1	-5,360762	28,7	
0,00068	16	14		5,8828	4,36076	0	0	0	34,6		5	25	21,803809	19	
0,0007	16	19		5,8828	9,36076	-1	1	-5,88279819	34,6		-4	16	-37,44305	87,6	
0,00072	15	15		4,8828	5,36076	2	4	9,76559639	23,8		-2	4	-10,72152	28,7	
0,00074	17	13		6,8828	3,36076	0	0	0	47,4		3	9	10,082285	11,3	
0,00076	17	16		6,8828	6,36076	0	0	0	47,4		-5	25	-31,80381	40,5	
0,00078	17	11		6,8828	1,36076	-1	1	-6,88279819	47,4		4	16	5,4430472	1,85	
0,0008	16	15		5,8828	5,36076	1	1	5,88279819	34,6		-1	1	-5,360762	28,7	
0,00082	17	14		6,8828	4,36076	-1	1	-6,88279819	47,4		1	1	4,3607618	19	
0,00084	16	15		5,8828	5,36076	-1	1	-5,88279819	34,6		0	0	0	28,7	
0,00086	15	15		4,8828	5,36076	-1	1	-4,88279819	23,8		1	1	5,3607618	28,7	
0,00088	14	16		3,8828	6,36076	2	4	7,76559639	15,1		1	1	6,3607618	40,5	
0,0009	16	17		5,8828	7,36076	-2	4	-11,7655964	34,6		-1	1	-7,360762	54,2	
0,00092	14	16		3,8828	6,36076	0	0	0	15,1		-2	4	-12,72152	40,5	
0,00094	14	14		3,8828	4,36076	0	0	0	15,1		0	0	0	19	
0,00096	14	14		3,8828	4,36076	-4	16	-15,5311928	15,1		-3	9	-13,08229	19	
0,00098	10	11		-0,1172	1,36076	1	1	-0,11720181	0,01		-1	1	-1,360762	1,85	
0,001	11	10		0,8828	0,36076	-1	1	-0,88279819	0,78		1	1	0,3607618	0,13	

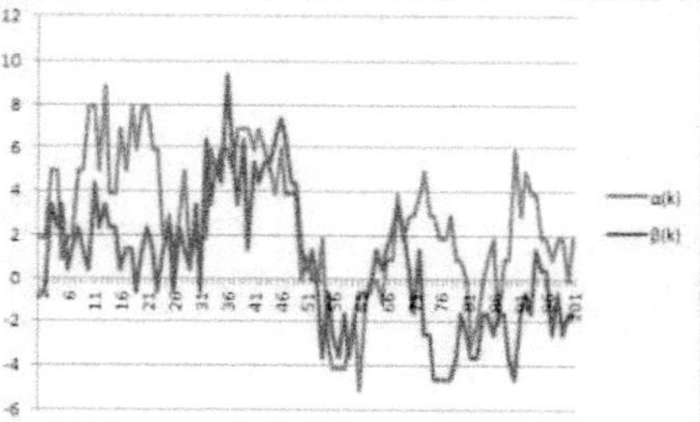

Figure 4.17. *Centered DMD $\alpha_k - \rho$: calculation of parameters σ_0, V, W, C and σ for the proportion of the mixture* $(0.0|1.0)$

The dependence of parameters on the mixture proportion allows us to obtain an unexpected result: the parameters σ and V have almost linear ability to distinguish the proportion of mixing of two ensembles of particles with different diffusion coefficients according to the Stokes–Einstein diffusion

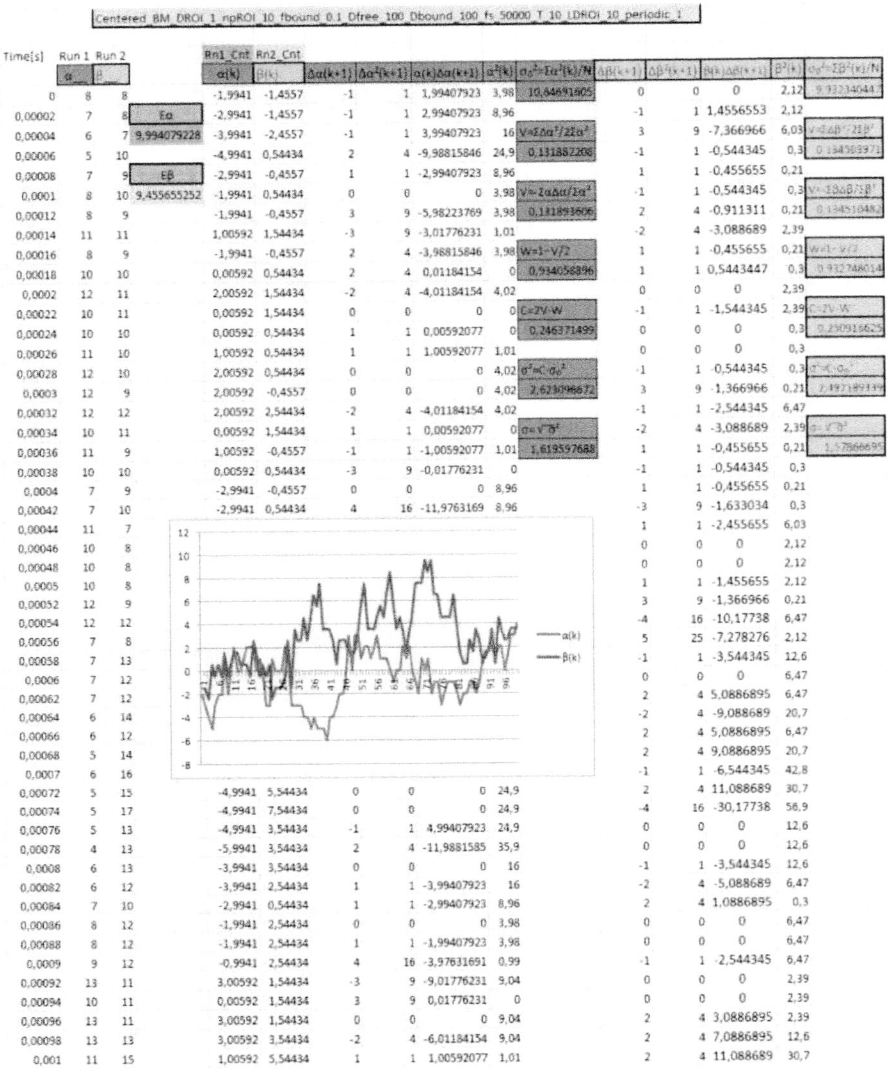

Figure 4.18. *Centered DMD $\alpha_k - \rho$: calculation of parameters σ_0, V, W, C and σ for the proportion of the mixture* $(0.1|0.9)$

model. However, the calculation of parameter σ_0^2 (diffusion coefficient of DMD) shows a significant intersection of the values σ_0^2 and means the absence of the property of parameter σ_0^2 to distinguish the mixing proportions of two ensembles of particles with different diffusion coefficients.

Centered_BM_DROI_1_npROI_10_fbound_0.1_Dfree_100_Dbound_100_fs_50000_T_10_LDROI_10_periodic_1

Time[s]	Run 1	Run 2		Rn1_Cnt	Rn2_Cnt										
	α	β		α(k)	β(k)	Δα(k+1)	Δα²(k+1)	α(k)Δα(k+1)	α²(k)	$\sigma_0^2=\Sigma\alpha^2(k)/N$	Δβ(k+1)	Δβ²(k+1)	β(k)Δβ(k+1)	β²(k)	$\sigma_0^2=\Sigma\beta^2(k)/N$
0	9	14		-0,790026247	4,031801257	0	0	0	0,62	9,806978151	0	0	0	16,3	8,527158124
0,00002	9	14	Eα	-0,790026247	4,031801257	0	0	0	0,62		0	0	0	16,3	
0,00004	9	14	9,790026247	-0,790026247	4,031801257	-2	4	1,58005249	0,62	$V=\Sigma\Delta\alpha^2/2\Sigma\alpha^2$	1	1	4,0318013	16,3	$V=\Sigma\Delta\beta^2/2\Sigma\beta^2$
0,00006	7	15		-2,790026247	5,031801257	1	1	-2,79002625	7,78	0,119755827	-1	1	-5,031801	25,3	0,142811849
0,00008	8	14	Eβ	-1,790026247	4,031801257	0	0	0	3,2		0	0	0	16,3	
0,0001	8	14	9,968198743	-1,790026247	4,031801257	-1	1	1,79002625	3,2	$V=\Sigma\alpha\Delta\alpha/\Sigma\alpha^2$	0	0	0	16,3	$V=\Sigma\beta\Delta\beta/\Sigma\beta^2$
0,00012	7	14		-2,790026247	4,031801257	0	0	0	7,78	0,119757769	-1	1	-4,031801	16,3	0,142670029
0,00014	7	13		-2,790026247	3,031801257	0	0	0	7,78		-1	1	-3,031801	9,19	
0,00016	7	12		-2,790026247	2,031801257	1	1	-2,79002625	7,78	W=1-V/2	2	4	4,0636025	4,13	W=1-V/2
0,00018	8	14		-1,790026247	4,031801257	-1	1	1,79002625	3,2	0,940122086	-2	4	-8,063603	16,3	0,928694075
0,0002	7	12		-2,790026247	2,031801257	1	1	-2,79002625	7,78		2	4	4,0636025	4,13	
0,00022	8	14		-1,790026247	4,031801257	1	1	-1,79002625	3,2	C=2V-W	0	0	0	16,3	C=2V-W
0,00024	9	14		-0,790026247	4,031801257	-2	4	1,58005249	0,62	0,225170196	-1	1	-4,031801	16,3	0,264885559
0,00026	7	13		-2,790026247	3,031801257	0	0	0	7,78		-1	1	-3,031801	9,19	
0,00028	7	12		-2,790026247	2,031801257	-1	1	2,79002625	7,78	$\sigma^2=C\cdot\sigma_0^2$	1	1	2,0318013	4,13	$\sigma^2=C\cdot\sigma_0^2$
0,0003	6	13		-3,790026247	3,031801257	-2	4	7,58005249	14,4	2,208238741	-1	1	-3,031801	9,19	2,258721049
0,00032	4	12		-5,790026247	2,031801257	-1	1	5,79002625	33,5		1	1	2,0318013	4,13	
0,00034	3	13		-6,790026247	3,031801257	0	0	0	46,1	$\sigma=\sqrt{\sigma^2}$	0	0	0	9,19	$\sigma=\sqrt{\sigma^2}$
0,00036	3	13		-6,790026247	3,031801257	1	1	-6,79002625	46,1	1,486014381	1	1	3,0318013	9,19	1,502904205
0,00038	4	14		-5,790026247	4,031801257	-3	9	17,3700787	33,5		0	0	0	16,3	
0,0004	1	14		-8,790026247	4,031801257	1	1	-8,79002625	77,3		0	0	0	16,3	
0,00042	2	14		-7,790026247	4,031801257	1	1	-7,79002625	60,7		-3	9	-12,0954	16,3	
0,00044	3	11									-1	1	-1,031801	1,06	
0,00046	3	10									-1	1	-0,031801	0	
0,00048	5	9									1	1	-0,968199	0,94	
0,0005	2	10									1	1	0,0318013	0	
0,00052	3	11									-2	4	-2,063603	1,06	
0,00054	4	9									1	1	-0,968199	0,94	
0,00056	2	10									-1	1	-0,031801	0	
0,00058	3	9									1	1	-0,968199	0,94	
0,0006	1	10									-1	1	-0,031801	0	
0,00062	2	9									4	16	-3,872795	0,94	
0,00064	1	13									-2	4	-6,063603	9,19	
0,00066	1	11									1	1	1,0318013	1,06	
0,00068	2	12									2	4	4,0636025	4,13	
0,0007	4	14									-1	1	-4,031801	16,3	
0,00072	5	13		-4,790026247	3,031801257	-1	1	4,79002625	22,9		1	1	3,0318013	9,19	
0,00074	4	14		-5,790026247	4,031801257	-1	1	5,79002625	33,5		-1	1	-4,031801	16,3	
0,00076	3	13		-6,790026247	3,031801257	1	1	-6,79002625	46,1		2	4	6,0636025	9,19	
0,00078	4	15		-5,790026247	5,031801257	0	0	0	33,5		-1	1	-5,031801	25,3	
0,0008	4	14		-5,790026247	4,031801257	0	0	0	33,5		0	0	0	16,3	
0,00082	4	14		-5,790026247	4,031801257	0	0	0	33,5		-1	1	-4,031801	16,3	
0,00084	4	13		-5,790026247	3,031801257	0	0	0	33,5		0	0	0	9,19	
0,00086	4	13		-5,790026247	3,031801257	0	0	0	33,5		-1	1	-3,031801	9,19	
0,00088	4	12		-5,790026247	2,031801257	1	1	-5,79002625	33,5		3	9	6,0954038	4,13	
0,0009	5	15		-4,790026247	5,031801257	0	0	0	22,9		2	4	10,063603	25,3	
0,00092	5	17		-4,790026247	7,031801257	-1	1	4,79002625	22,9		-3	9	-21,0954	49,4	
0,00094	4	14		-5,790026247	4,031801257	0	0	0	33,5		-1	1	-4,031801	16,3	
0,00096	4	13		-5,790026247	3,031801257	0	0	0	33,5		1	1	3,0318013	9,19	
0,00098	4	14		-5,790026247	4,031801257	0	0	0	33,5		1	1	4,0318013	16,3	
0,001	4	15		-5,790026247	5,031801257	0	0	0	33,5		0	0	0	25,3	

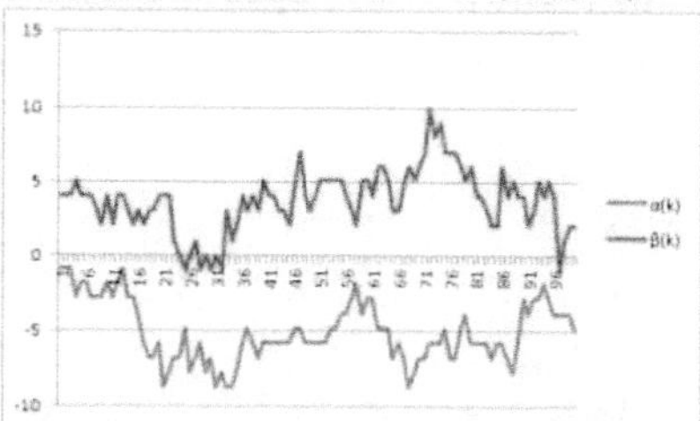

Figure 4.19. *Centered DMD $\alpha_k - \rho$: calculation of parameters σ_0, V, W, C and σ for the proportion of the mixture* $(0.3|0.7)$

Conclusions on DMD simulation results in biophysics

We can summarize the comparative analysis with the simulation results as follows:

1) For a mathematical description of the interaction of macromolecules, a statistical model is presented by the solution of equation [4.2.1], characterized by numerical parameters V (regression of the predicted component) and σ^2 (the square variation of the stochastic component).

Centered_BM_DROI_1_npROI_10_fbound_0.7_Dfree_100_Dbound_100_fs_50000_T_10_LDROI_10_periodic_1

Time[s]	Run 1	Run 2		Rn1_Cnt	Rn2_Cnt										
	α	β		α(k)	β(k)	Δα(k+1)	Δα²(k+1)	α(k)Δα(k+1)	α²(k)	σ0²=Σα²(k)/N	Δβ(k+1)	Δβ²(k+1)	β(k)Δβ(k+1)	β²(k)	σ0²=Σβ²(k)/N
0	9	9		-0,8465	-0,2325	2	4	-1,6930965	0,72	7,807985247	4	16	-0,929988	0,05	9,008204758
0,00002	11	13	Eα	1,15345	3,7675	-1	1	-1,15345175	1,33		-2	4	-7,535006	14,2	
0,00004	10	11	9,846548251	0,15345	1,7675	0	0	0	0,02	V=ΣΔα²/2Σα²	1	1	1,7675029	3,12	V=ΣΔβ²/2Σβ²
0,00006	10	12		0,15345	2,7675	1	1	0,15345175	0,02	0,097673958	0	0	0	7,66	0,088277351
0,00008	11	12	Eβ	1,15345	2,7675	0	0	0	1,33		-2	4	-5,535006	7,66	
0,0001	11	10	9,232497101	1,15345	0,7675	1	1	1,15345175	1,33	V=-ΣαΔα/Σα²	-1	1	-0,767503	0,59	V=-ΣβΔβ/Σβ²
0,00012	12	9		2,15345	-0,2325	0	0	0	4,64	0,097676759	2	4	-0,464994	0,05	0,088277537
0,00014	12	11		2,15345	1,7675	0	0	0	4,64		2	4	3,5350058	3,12	
0,00016	12	13		2,15345	3,7675	0	0	0	4,64	W=1-V/2	-1	1	-3,767503	14,2	W=1-V/2
0,00018	12	12		2,15345	2,7675	-2	4	-4,3069035	4,64	0,951163021	-1	1	-2,767503	7,66	0,955861323
0,0002	10	11		0,15345	1,7675	1	1	0,15345175	0,02		0	0	0	3,12	
0,00022	11	11		1,15345	1,7675	-1	1	-1,15345175	1,33	C=2V·W	0	0	0	3,12	C=2V·W
0,00024	10	11		0,15345	1,7675	0	0	0	0,02	0,185807713	-1	1	-1,767503	3,12	0,168761816
0,00026	10	10		0,15345	0,7675	-1	1	-0,15345175	0,02		1	1	0,7675029	0,59	
0,00028	9	11		-0,8465	1,7675	1	1	-0,84654825	0,72	σ²=C·σ0²	-1	1	-1,767503	3,12	σ²=C·σ0²
0,0003	10	10		0,15345	0,7675	0	0	0	0,02	1,450783885	2	4	1,5350058	0,59	1,520240991
0,00032	10	12		0,15345	2,7675	-1	1	-0,15345175	0,02		2	4	5,5350058	7,66	
0,00034	9	14		-0,8465	4,7675	-2	4	1,6930965	0,72	σ=√σ²	-3	9	-14,30251	22,7	σ=√σ²
0,00036	7	11		-2,8465	1,7675	0	0	0	8,1	1,204484904	0	0	0	3,12	1,232980531
0,00038	7	11		-2,8465	1,7675	1	1	-2,84654825	8,1		0	0	0	3,12	
0,0004	8	11		-1,8465	1,7675	0	0	0	3,41		2	4	3,5350058	3,12	
0,00042	8	13		-1,8465	3,7675	0	0	0	3,41		0	0	0	14,2	
0,00044	8	13									1	1	3,7675029	14.2	
0,00046	7	14									0	0	0	22,7	
0,00048	7	14									0	0	0	22,7	
0,0005	7	14									0	0	0	22,7	
0,00052	8	14									1	1	4,7675029	22,7	
0,00054	5	15									0	0	0	33,3	
0,00056	6	15									1	1	5,7675029	33,3	
0,00058	6	16									-2	4	-13,53501	45,8	
0,0006	6	14									1	1	4,7675029	22,7	
0,00062	6	15									0	0	0	33,3	
0,00064	7	15									-2	4	-11,53501	33,3	
0,00066	6	13									-3	9	-11,30251	14,2	
0,00068	6	10									1	1	0,7675029	0,59	
0,0007	6	11									3	9	5,3025087	3,12	
0,00072	6	14		-3,8465	4,7675	0	0	0	14,8		2	4	9,5350058	22,7	
0,00074	6	16		-3,8465	6,7675	1	1	-3,84654825	14,8		2	4	13,535006	45,8	
0,00076	7	18		-2,8465	8,7675	0	0	0	8,1		-1	1	-8,767503	76,9	
0,00078	7	17		-2,8465	7,7675	0	0	0	8,1		-1	1	-7,767503	60,3	
0,0008	7	16		-2,8465	6,7675	0	0	0	8,1		-3	9	-20,30251	45,8	
0,00082	7	13		-2,8465	3,7675	1	1	-2,84654825	8,1		-1	1	-3,767503	14,2	
0,00084	8	12		-1,8465	2,7675	-1	1	1,84654825	3,41		-1	1	-2,767503	7,66	
0,00086	7	11		-2,8465	1,7675	0	0	0	8,1		0	0	0	3,12	
0,00088	7	11		-2,8465	1,7675	0	0	0	8,1		0	0	0	3,12	
0,0009	7	11		-2,8465	1,7675	1	1	-2,84654825	8,1		0	0	0	3,12	
0,00092	8	11		-1,8465	1,7675	-2	4	3,6930965	3,41		-1	1	-1,767503	3,12	
0,00094	6	10		-3,8465	0,7675	1	1	-3,84654825	14,8		-2	4	-1,535006	0,59	
0,00096	7	8		-2,8465	-1,2325	0	0	0	8,1		0	0	0	1,52	
0,00098	7	8		-2,8465	-1,2325	1	1	-2,84654825	8,1		1	1	-1,232497	1,52	
0,001	8	9		-1,8465	-0,2325	1	1	-1,84654825	3,41		0	0	0	0,05	

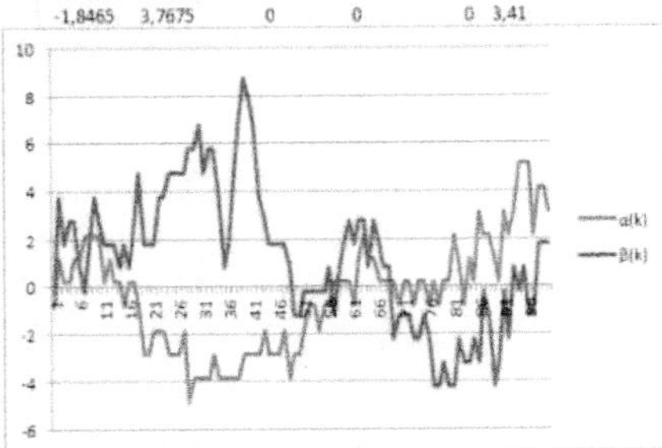

Figure 4.20. *Centered DMD $\alpha_k - \rho$: calculation of parameters σ_0, V, W, C and σ for the proportion of the mixture* $(0.7|0.3)$

2) The obtained statistical estimates [4.3.4]–[4.3.6] of the interaction parameters $(V,\ \sigma^2)$, which have a convenient form for computation, in comparison with the traditional method of fitting the autocorrelation curves based on the kinetic Stokes–Einstein model [4.3.11] and [4.3.12].

3) The statistical estimates [4.3.4]–[4.3.6] apply to numerical modeling of the process determined by the statistical model [4.2.1], and there is a convergence of estimation of parameters to their real value, which shows the adequacy of the estimates [4.3.4]–[4.3.6].

4) The proposed model was tested on the set of simulation data obtained from the discrete kinetic Stokes–Einstein Brownian process of two biological molecules with different diffusion coefficients and different molar fractions. Statistics [4.3.3]–[4.3.6] have a quasilinear discriminant dependence on the proportion of different molecular fractions of a mixture of molecules.

As a rule, biochemical systems are rather complex, with the participation of many types of interactions, and thus, it is difficult to study and analyze them. A new method for the study of complex interactions is proposed, which uses experimental techniques in conjunction with stochastic difference equations for extrapolation of parameters directly related to the dynamics of biochemical systems.

The proposed model can be used for a number of biochemical reaction models. A possible application in biophysics is the study of the biological diffusion of macromolecules in a solution (e.g., enzymes and substrates), combining our model with a set of data acquired through a system of fluorescence correlation spectroscopy (FCS). In this case, in fact, the fluctuations in the fluorescence signal of the labeled molecules are observed in a small (infinitesimal) volume of observations, which allows for a thorough study of the mechanisms of interactions.

The obtained statistical estimates [4.3.4]–[4.3.6] of the interaction parameters (V, σ^2) have more prospects for accuracy and efficiency than the traditional method of fitting the autocorrelation curves based on the Stokes–Einstein kinetic models [4.3.11] and [4.3.12].

Applying work based on the proposed model to the experimental data of FCS, it is possible to directly determine the diffusion coefficients and the molar fraction of the molecules under study. In particular, we can observe the dynamics of binding/solving enzyme macromolecules and their substrates, in accordance with the mass change of observed quantities. These processes can be of fundamental importance for the development of new drugs as inhibitors of artificial enzymes that are involved in pathologies.

References

Barndorff-Nielsen, O.E. and Shiryaev, A.N. (2010). *Change of Time and Change of Measure*. World Scientific, New York and London.

Borovkov, A.A. (1988). *Estadística Matemática*. Editorial Mir, Moscow.

Borovskikh, Y.V., and Korolyuk, V.S. (1997). *Martingale Approximation*. VSP, Utrecht.

Bousquet, O., Boucheron, S., and Lugosi, G. (2004). Introduction to statistical learning theory. In *Advanced Lectures on Machine Learning*, Springer-Verlag, Berlin and Heidelberg, pp. 169–207.

Bulinski, A. and Shiryaev, A.N. (2003). *Theory of Random Processes*. Fismatlit, Moscow.

Crow, J.F. and Kimura, M. (2009). *An Introduction to Population Genetics Theory*. Blackburn Press, Cardwell.

Ethier, S.N. and Kurtz, T.G. (1986). *Markov Processes: Characterization and Convergence*. Wiley, Hoboken.

Feng, J. and Kurtz, T.G. (2006). *Large Deviations for Stochastic Processes*. AMS, New York and Boston.

Fischer, H.P. (2008). Mathematical modeling of complex biological systems. *Alcohol Research and Health*, 31, 49–59.

Fisher, R.A. (1930). Genetics, mathematics, and natural selection. *Nature*, 126, 805–806.

Freidlin, M.I. and Ventzell, A.D. (2012). *Random Perturbation of Dynamical Systems*. Springer-Verlag, Berlin and Heidelberg.

Gihman, I.I. and Skorohod, A.V. (1979). *Controlled Stochastic Processes*. Springer-Verlag, New York.

Gnedenko, B.V. (2011). *A Course of Probability Theory*. Librocom, Moscow.

Heyde, C.C. (1997). *Quasi-Likelihood and its Application. A General Approach to Optimal Parameter Estimation*. Springer-Verlag, New York.

Ibragimov, I.A. and Has'minskii, R.Z. (1981). *Statistical Estimation. Asymptotic Theory*. Springer-Verlag, New York.

Jacod, J. and Shiryaev, A.N. (1987). *Limit Theorems for Stochastic Processes*. Springer-Verlag, Berlin and Heidelberg.

Keifer, E. and Wolfowitz, J. (1952). Stochastic estimation of the maximum of a regression function. *Annals of Mathematical Statistics*, 23(3), 462–466.

Koroliouk, D. (2015a). Binary statistical experiments with persistent nonlinear regression. *Theory of Probability and Mathematical Statistics*, 91, 71–80.

Koroliouk, D. (2015b). Binary statistical experiments with persistent linear regression in the Markov random medium. *Reports of the National Academy of Science of Ukraine*, 4, 12–17.

Koroliouk, D. (2015c). Two component binary statistical experiments with persistent linear regression. *Theory of Probability and Mathematical Statistics*, 90, 103–114.

Koroliouk, D. (2016a). Multivariate statistical experiments with persistent linear regression and equilibrium. *Theory of Probability and Mathematical Statistics*, 92, 71–79.

Koroliouk, D. (2016b). Stationary statistical experiments and the optimal estimator for a predictable component. *Journal of Mathematical Sciences*, 214(2), 220–228.

Koroliouk, D. (2016c). The problem of discrete Markov diffusion leaving an interval. *Cybernetics and Systems Analytics*, 52(4), 571–576.

Koroliouk, D. (2017). Adapted statistical experiments. *Journal of Mathematical Sciences*, 220(5), 615–623.

Koroliouk, D. and Koroliuk, V.S. (2019). Equilibrium in Wright-Fisher models of population genetics. *Cybernetics and Systems Analytics*, 55(2), 253–258.

Koroliouk, D., Koroliuk, V.S., and Rosato, N. (2014). Equilibrium process in biomedical data analysis: the Wright-Fisher model. *Cybernetics and Systems Analysis*, 50(6), 890–897.

Koroliouk, D., Bertotti, M.L., and Koroliuk, V. S. (2016a). Stochastic behavioral models. Classification. *Cybernetics and Systems Analysis*, 52(6), 884–895.

Koroliouk, D., Koroliuk, V.S., Nicolai, E., Bisegna, P., Stella, L., and Rosato, N. (2016b). A statistical model of macromolecules dynamics for Fluorescence Correlation Spectroscopy data analysis. *Statistics, Optimization and Information Computing*, 4, 233–242.

Korolyuk, V.S. and Korolyuk, V.V. (1999). *Stochastic Models of Systems*. Kluwer, Dodrecht and Boston.

Korolyuk, V.S. and Limnios, N. (2005). *Stochastic Systems in Merging Phase Space*. World Scientific, New Jersey and London.

Krasovskii, N.N. (1963). *Stability of Motion. Applications of Lyapunov's Second Method*. Stanford University Press, Stanford.

Limnios, N. and Samoilenko, I. (2013). Poisson approximation of processes with locally independent increments with Markov switching. *Teor. Imovir. ta Matem. Statyst.*, 89, 104–114.

Liggett, T.M. (2010). *Continuous Time Markov Processes: An Introduction* (vol. 113). AMS, Providence.

Liptser, R.S. (1994). The Bogolyubov averaging principle for semimartingales. *Proceedings of the Steklov Institute of Mathematics—Moscow*, 4, 1–12.

Liptser, R.S. and Shiryaev, A.N. (2001). *Statistics of Random Processes. II. Applications*. Springer-Verlag, Berlin and Heidelberg.

Moklyachuk, M.P. (1994). *Stochastic autoregressive sequences and minimax interpolation. Theory of Probability and Mathematical Statistics*, 48, 95–104.

Nevelson, M.B. and Has'minskii, R.Z. (1976). *Stochastic Approximation and Recursive Estimation*. American Mathematical Society, Providence.

Rigler, R. and Elson, E.S. (2001). *Fluorescence Correlation Spectroscopy. Theory and Applications*. Springer-Verlag, Berlin and Heidelberg.

Robbins, H. and Monro, S. (1951). A stochastic approximation method. *Annals of Mathematical Statistics*, 22(1), 400–407.

Schoen, R. (2006). *Dynamic Population Models*. Springer-Verlag, Dordrecht.

Schwille, P., Meyer-Almes, F.J., and Rigler, R. (1997). Dual-color fluorescence cross-correlation spectroscopy for multicomponent diffusional analysis in solution. *Biophysical Journal*, 72(4), 1878–1886.

Shiryaev, A.N. (1999). *Essentials of Stochastic Finance: Facts, Models, Theory*. World Scientific, New York.

Shiryaev, A.N. (2011). *Probability* (vol. 1). Springer-Verlag, Berlin.

Skorokhod, A.V. (1987). *Asymptotic Methods in the Theory of Stochastic Differential Equations*. Naukova Dumka, Kiev (in Russian).

Skorokhod, A.V., Hoppensteadt, F.C., and Salehi, H. (2002). *Random Perturbation Methods with Applications in Science and Engineering*. Springer, New York.

Svirezhev, Y.M. and Pasekov, V.P. (1982). *Fundamentals of Mathematical Genetics*. Nauka, Moscow (in Russian).

Vapnik, V.N. (2000). *The Nature of Statistical Learning Theory*. Springer-Verlag, New York.

Wright, S. (1969). *Evoution and the Genetics of Populations, V. 2. The Theory of Gene Frequencies*. University of Chicago Press, Chicago.

Index

A, B, C

D

E, F

G, H, K, L

M, N

O, P, Q

R, S

T, W

Other titles from

iSTE

in

Mathematics and Statistics

2019

BANNA Oksana, MISHURA Yuliya, RALCHENKO Kostiantyn, SHKLYAR Sergiy
Fractional Brownian Motion: Approximations and Projections

GANA Kamel, BROC Guillaume
Structural Equation Modeling with lavaan

KUKUSH Alexander
Gaussian Measures in Hilbert Space: Construction and Properties

LUZ Maksym, MOKLYACHUK Mikhail
Estimation of Stochastic Processes with Stationary Increments and Cointegrated Sequences

MICHELITSCH Thomas, PÉREZ RIASCOS Alejandro, COLLET Bernard, NOWAKOWSKI Andrzej, NICOLLEAU Franck
Fractional Dynamics on Networks and Lattices

VOTSI Irene, LIMNIOS Nikolaos, PAPADIMITRIOU Eleftheria, TSAKLIDIS George
Earthquake Statistical Analysis through Multi-state Modeling (Statistical Methods for Earthquakes Set – Volume 2)

2018

AZAÏS Romain, BOUGUET Florian
Statistical Inference for Piecewise-deterministic Markov Processes

IBRAHIMI Mohammed
Mergers & Acquisitions: Theory, Strategy, Finance

PARROCHIA Daniel
Mathematics and Philosophy

2017

CARONI Chysseis
First Hitting Time Regression Models: Lifetime Data Analysis Based on Underlying Stochastic Processes
(Mathematical Models and Methods in Reliability Set – Volume 4)

CELANT Giorgio, BRONIATOWSKI Michel
Interpolation and Extrapolation Optimal Designs 2: Finite Dimensional General Models

CONSOLE Rodolfo, MURRU Maura, FALCONE Giuseppe
Earthquake Occurrence: Short- and Long-term Models and their Validation
(Statistical Methods for Earthquakes Set – Volume 1)

D'AMICO Guglielmo, DI BIASE Giuseppe, JANSSEN Jacques, MANCA Raimondo
Semi-Markov Migration Models for Credit Risk
(Stochastic Models for Insurance Set – Volume 1)

GONZÁLEZ VELASCO Miguel, del PUERTO GARCÍA Inés, YANEV George P.
Controlled Branching Processes
(Branching Processes, Branching Random Walks and Branching Particle Fields Set – Volume 2)

HARLAMOV Boris
Stochastic Analysis of Risk and Management
(Stochastic Models in Survival Analysis and Reliability Set – Volume 2)

KERSTING Götz, VATUTIN Vladimir
Discrete Time Branching Processes in Random Environment
(Branching Processes, Branching Random Walks and Branching Particle Fields Set – Volume 1)

MISHURA YULIYA, SHEVCHENKO Georgiy
Theory and Statistical Applications of Stochastic Processes

NIKULIN Mikhail, CHIMITOVA Ekaterina
Chi-squared Goodness-of-fit Tests for Censored Data
(Stochastic Models in Survival Analysis and Reliability Set – Volume 3)

SIMON Jacques
Banach, Fréchet, Hilbert and Neumann Spaces
(Analysis for PDEs Set – Volume 1)

2016

CELANT Giorgio, BRONIATOWSKI Michel
Interpolation and Extrapolation Optimal Designs 1: Polynomial Regression and Approximation Theory

CHIASSERINI Carla Fabiana, GRIBAUDO Marco, MANINI Daniele
Analytical Modeling of Wireless Communication Systems
(Stochastic Models in Computer Science and Telecommunication Networks Set – Volume 1)

GOUDON Thierry
Mathematics for Modeling and Scientific Computing

KAHLE Waltraud, MERCIER Sophie, PAROISSIN Christian
Degradation Processes in Reliability
(Mathematial Models and Methods in Reliability Set – Volume 3)

KERN Michel
Numerical Methods for Inverse Problems

RYKOV Vladimir
Reliability of Engineering Systems and Technological Risks
(Stochastic Models in Survival Analysis and Reliability Set – Volume 1)

2015

DE SAPORTA Benoîte, DUFOUR François, ZHANG Huilong
Numerical Methods for Simulation and Optimization of Piecewise Deterministic Markov Processes

DEVOLDER Pierre, JANSSEN Jacques, MANCA Raimondo
Basic Stochastic Processes

LE GAT Yves
Recurrent Event Modeling Based on the Yule Process
(Mathematical Models and Methods in Reliability Set – Volume 2)

2014

COOKE Roger M., NIEBOER Daan, MISIEWICZ Jolanta
Fat-tailed Distributions: Data, Diagnostics and Dependence
(Mathematical Models and Methods in Reliability Set – Volume 1)

MACKEVIČIUS Vigirdas
Integral and Measure: From Rather Simple to Rather Complex

PASCHOS Vangelis Th
Combinatorial Optimization – 3-volume series – 2nd edition
Concepts of Combinatorial Optimization / Concepts and Fundamentals – volume 1
Paradigms of Combinatorial Optimization – volume 2
Applications of Combinatorial Optimization – volume 3

2013

COUALLIER Vincent, GERVILLE-RÉACHE Léo, HUBER Catherine, LIMNIOS Nikolaos, MESBAH Mounir
Statistical Models and Methods for Reliability and Survival Analysis

JANSSEN Jacques, MANCA Oronzio, MANCA Raimondo
Applied Diffusion Processes from Engineering to Finance

SERICOLA Bruno
Markov Chains: Theory, Algorithms and Applications

2012

BOSQ Denis
Mathematical Statistics and Stochastic Processes

CHRISTENSEN Karl Bang, KREINER Svend, MESBAH Mounir
Rasch Models in Health

DEVOLDER Pierre, JANSSEN Jacques, MANCA Raimondo
Stochastic Methods for Pension Funds

2011

MACKEVIČIUS Vigirdas
Introduction to Stochastic Analysis: Integrals and Differential Equations

MAHJOUB Ridha

Recent Progress in Combinatorial Optimization – ISCO2010

RAYNAUD Hervé, ARROW Kenneth
Managerial Logic

2010

BAGDONAVIČIUS Vilijandas, KRUOPIS Julius, NIKULIN Mikhail
Nonparametric Tests for Censored Data

BAGDONAVIČIUS Vilijandas, KRUOPIS Julius, NIKULIN Mikhail
Nonparametric Tests for Complete Data

IOSIFESCU Marius *et al.*
Introduction to Stochastic Models

VASSILIOU PCG
Discrete-time Asset Pricing Models in Applied Stochastic Finance

2008

ANISIMOV Vladimir
Switching Processes in Queuing Models

FICHE Georges, HÉBUTERNE Gérard
Mathematics for Engineers

HUBER Catherine, LIMNIOS Nikolaos *et al.*
Mathematical Methods in Survival Analysis, Reliability and Quality of Life

JANSSEN Jacques, MANCA Raimondo, VOLPE Ernesto
Mathematical Finance

2007

HARLAMOV Boris
Continuous Semi-Markov Processes

2006

CLERC Maurice
Particle Swarm Optimization

Printed and bound by CPI Group (UK) Ltd, Croydon, CR0 4YY